Lecture Notes in Mathematics

Edited by A. Dold and B. Eckmann

Subseries: USSR

Adviser: L. D. Faddeev, Leningrad

1191

Alexander R. Its
Victor Yu. Novokshenov

The Isomonodromic Deformation Method in the Theory of Painlevé Equations

Springer-Verlag
Berlin Heidelberg New York Tokyo

Authors

Alexander R. Its
Leningrad State University, Department of Physics
St. Peterhoff, 198 904 Leningrad, USSR

Victor Yu. Novokshenov
Bashkir Branch of the Academy of Sciences of the USSR
Tukaeva 50, 450057 Ufa, USSR

Consulting Editor

Leon A. Takhtajan
LOMI
Fontanka 27, Leningrad, 191011, USSR

Mathematics Subject Classification (1980): 34A34, 35Q20

ISBN 3-540-16483-9 Springer-Verlag Berlin Heidelberg New York Tokyo
ISBN 0-387-16483-9 Springer-Verlag New York Heidelberg Berlin Tokyo

Printed in Germany

Printing and binding: Beltz Offsetdruck, Hemsbach/Bergstr.
2146/3140-543210

CONTENTS

Chapter 0. INTRODUCTION

The Painlevé equations appeared in the theory of ordinary differential equations at the beginning of our century in connection with a classification problem for the equations of the form

$$u_{xx} = R(x, u, u_x). \tag{0.1}$$

The function R here is supposed to be analytic in x and rational in u and u_x. Under these conditions the general integral u of equation (0.1) must have no movable (i.e. depending on initial data) branch-type singularities (the so-called Painlevé property).

The classification problem was solved completely by P.Painlevé and B.Gambier [1], [2]. They discovered just 50 canonical types of equations of the form (0.1) (up to local transformations admitted by the right-hand side) posessing the Painlevé-property. It turns out that only six of them can not be reduced to linear equations, i.e. their general integral is not expressed in terms of known special functions. These six equations are called the Painlevé equations (PI -PVI) and corresponding solutions - the Painlevé functions or Painlevé transcendents. The complete list of these equations is given in Appendix 3 of the present paper.

Further studies of the ordinary differential equations possessing the Painlevé-property were concentrated on the investigations of qualitative properties of their solutions, distribution of their movable poles, finding out the conditions providing the existence of rational solutions or the special function solutions and the construction of different procedures producing a new solution starting from the known one. All these problems were considered in detail in papers by N.P.Erugin, N.L.Lukashevich, A.I.Yablonsky, V.I.Gromak and others by means of conventional methods of the analytical theory of differential equations (see the reviews by N.P.Erugin [3] (1976) and by L.A.Bordag [57] (1980)).

The results of their investigations were further developed in connection with the discovery of the inverse scattering method in the theory of non-linear equations. A great number of papers cencerning particular solutions of Painlevé equations has since appeared. We, however, will not consider this question and so restrict ourselves to referring to the paper by A.S.Focas and M.J.Ablowitz [4], where a systematical approach to construction of the Bäcklund transforms for all the six Painlevé equations are suggested and the main results mentioned above are also reviewed.

A new surge of interest in Painlevé equations is due to their appearence in concrete problems of theoretical and mathematical physics. They happened to be closely connected with the quantum field theory [5] - [11] and the non-linear evolutionary equations [11] - [20]. The characteristic feature of Painlevé equations in this context is that they describe certain transitional and self-similar processes. Furthermore, according to [19], [20] it does not matter whether the initial system is completely integrable or not. In other words the recent developments in non-linear theoretical physics draw one to the conclusion that Painlevé transcendents play just the same role as classical special functions in linear problems.

The analytic basis for the wide use of special functions in mathematical physics is essentially provided by the existence of explicit formulae linking their asymptotics at different characteristical points. This fact is based on the possibility of application of the Laplace's method to linear differential equations associated with classical special functions. It should be emphasized that the non-linear analogue of the Laplace's method, the so called isomonodromic deformation method (IDM), was recently found by H.Flashka and A.Newell [16], M.Jimbo, T.Miwa and K.Ueno [14]. This fact is a justification of the analogy mentioned above between the Painlevé transcendents and the classical special functions. The IDM (as a Laplace's method for a linear theory) permits one to obtain the explicit connection formulae for the solutions of Painlevé equations.

Let us sketch briefly the essence of isomonodromic deformation method. The idea is to associate with initial non-linear differential equation of the form (0.1) a certain linear system

$$\frac{d\Psi}{d\lambda} = A(\lambda, x, u, u_x)\Psi \tag{0.2}$$

with matrix-valued coefficients rational in λ . The deformations of coefficients in x is described by equation (0.1) in such a way that the monodromy data of system (0.2) *) have been conserved. For example, Painlevé II equation

$$u_{xx} - xu - 2u^3 = \nu \tag{0.3}$$

is associated with the system ([16])

$$\frac{d\Psi}{d\lambda} = \begin{pmatrix} -4i\lambda^2 - ix - 2iu^2, & 4i\lambda u - 2u_x - \frac{i\nu}{\lambda} \\ -4i\lambda u - 2u_x + \frac{i\nu}{\lambda}, & 4i\lambda^2 + ix + 2iu^2 \end{pmatrix} \Psi . \tag{0.4}$$

Therefore the monodromy data of the linear equation (0.2) present the first integrals of non-linear differential equation (0.1). So the problem of integration of this equation is reduced to classical problem

---

*) Strictly speaking, the idea to represent the Painlevé equations as an isomonodromic deformation condition for certain linear differential equation with rational coefficients ascents to the works of R.Fuchs [21] and R.Garnier [22] . Nevertheless we associate the appearance of isomonodromic deformation method rather with the recent works [16] , [14] than with the classical ones. The reasons are explained at the end of the Introduction.

of linear analysis - the solution of direct and inverse problems of monodromy theory for the systems of linear equations with rational coefficients. We outline the complete analogy with the inverse scattering method in the theory of non-linear partial differential equations, where the integration procedure is reduced to the solution of direct and inverse spectral problems for suitable linear differential operator (U-operator). Moreover, the IDM itself appears to be to some extent the corollary of the inverse scattering method. As we have already mentioned above the Painlevé equations describe self-similar solutions of non-linear evolutionary equations being integrable by the inverse scattering method. The corresponding "equation in λ " (0.2) associated with Painlevé equation might be deduced from the initial $U-V$ pair under suitable self-similar reduction. We illustrate this procedure (following [16]) using the Painlevé II equation as an example. Assume $u(x)$ to be the solution to equation (0.3) and put

$$y(\eta, t) = (3t)^{-\frac{1}{3}} u(\eta \cdot (3t)^{-\frac{1}{3}}) ,$$

which represents self-similar solution of modified Korteweg-de Vries equation (MKdV)

$$y_t - 6y^2 y_\eta + y_{\eta\eta\eta} = 0. \tag{0.5}$$

The two linear systems with the compatibility condition presented by equation (0.5), is the well-known $U-V$ pair for MKdV:

$$\frac{d\Phi}{d\eta} = \begin{pmatrix} -iz & iy \\ -iy & iz \end{pmatrix} \Phi \equiv U(z)\Phi , \tag{0.6}$$

$$\frac{d\Phi}{dt} = \begin{pmatrix} -4iz^3 - 2iy^2 z \;, & 4iz^2 y - 2zy_\eta - iy_{\eta\eta} + 2iy^3 \\ -4iz^2 y - 2zy_\eta + iy_{\eta\eta} - 2iy^3 \;, & 4iz^3 + 2iy^2 z \end{pmatrix} \Phi \equiv$$

$$\equiv V(z)\,\Phi$$

Let us rescale here the variables, putting

$$\begin{aligned} &\eta \longmapsto x = \eta \cdot (3t)^{-\frac{1}{3}} , \\ &z \longmapsto \lambda = z \cdot (3t)^{-\frac{1}{3}} , \\ &y \longmapsto u, \quad y(\eta, t) = (3t)^{-\frac{1}{3}} u(x)\Big|_{x=\eta\cdot(3t)^{-\frac{1}{3}}} , \\ &\Phi \longmapsto \Psi , \\ &\Phi(z,\eta,t) = \Psi(\lambda, x) \;\Big|\; \begin{array}{l} \lambda = z \cdot (3t)^{-1/3}, \\ x = \eta \cdot (3t)^{-1/3} . \end{array} \end{aligned} \tag{0.7}$$

Hence we have

$$\begin{aligned} &\frac{\partial \Phi}{\partial \eta} = \frac{\partial \Psi}{\partial x} \cdot (3t)^{-1/3} , \\ &\frac{\partial \Phi}{\partial t} = \frac{\partial \Psi}{\partial \lambda} \cdot \frac{\lambda}{3t} - \frac{\partial \Psi}{\partial x} \cdot \frac{x}{3t} , \\ &y_\eta = (3t)^{-2/3} u_x \,, \quad y_{\eta\eta} = \frac{1}{3t} u_{xx} . \end{aligned} \tag{0.8}$$

Substituting (0.7), (0.8) into the system (0.6) and applying (0.5) we transform (0.6) into the following system:

$$\frac{\partial \Psi}{\partial x} = \begin{pmatrix} -i\lambda \;, & iu \\ -iu \;, & i\lambda \end{pmatrix} \Psi , \tag{0.9}$$

$$\frac{\partial\Psi}{\partial\lambda} = \begin{pmatrix} -4i\lambda^2-2iu^2-ix, & 4i\lambda u-2u_x-\frac{i\nu}{\lambda} \\ -4i\lambda u-2u_x+\frac{i\nu}{\lambda}, & 4i\lambda^2+2iu^2+ix \end{pmatrix}\Psi . \tag{0.10}$$

The compatibility condition of (0.9), (0.10) (as we have known from the very beginning!) is just the equation (0.3). Note finally that equation (0.10) coincides with equation (0.4) and the additional equation (0.9) implies (see [14] , [16] and Chapters 2, 3 of the present paper) the fact that the monodromy data (Stokes matrices) for the system (0.10) do not depend on the deformation parameter x .

Some interesting feature of IDM is worth mentioning. As a result of our transformation of initial "spectral" $U-V$ pair into "isomonodromic" $A-U$ pair the role played by operators U, V has been essentially changed. The V operator being used only for the description of spectral data evolution of the operator $\frac{d}{dx} - U(\lambda)$ becomes the main object of the IDM. It represents now the "equation in λ ". On the other hand the U operator which played the basis role, transforms into an auxiliary one. It describes now "isomonodromoc" dynamics of solutions to the main A -equation (0.2). As a matter of fact one can forget about the U -equation in any concrete application and concentrate oneself on the studies of equation in λ . Therefore in Appendix 3, where we present (following [15]) the list of all Painlevé equations together with corresponding systems (0.2), the U -equations are omitted

The direct and inverse problems of the monodromy theory for the systems with rational coefficients being applied for the integration of Painlevé equations, are in fact essentially transcendent problems ([23] - [26]). Their explicit solution is available only for particular systems which may be reduced to hypergeometric equation. A question arises naturally about the effectiveness of IDM. We hope that the results obtained in the present paper are will contribute to the positive solution of this question. To demonstrate possibilities of the IDM we show how the central problems of Painlevé equations theory might be solved. Explicit calculations of asymptotics of solutions to the Cauchy problem, the construction of connection formulae for asymptotic in different domains and an explicit description of movable singularities distribution in terms of initial data are examples of these problems. An analytic basis for the achievement of these results consists of possibility of asymptotic solution to a direct problem of monodromy theory for the system (0.2) associated with Painlevé equations. It supposes a calculation of monodromy data through the prescribed values u, u_x under certain assumptions about an asymptotic behavior of the latter. Apriori information about their behaviour might be extracted independently of IDM by an application of qualitative analysis of solutions to Painlevé equation itself.

An asymptotic solution of the monodromy theory direct problem for the system (0.2) leads to the expression of the same integrals of equation (0.1) (the Stokes matrices of system (9.2)) in two different ways - through the asymptotic characteristics of the same solution $u(x)$ in two different domains of x ($x \sim +\infty$, $x \sim -\infty$ or $x \sim +\infty$, $x \sim 0$). This allows us to connect the characteristics in question by the explicit formulae, i.e. to solve a problem being earlier solved only for the classical special functions. For the first time as far as we know the program described above was accomplished by the second author [27] when calculating the asymptotics of a regular solution to the equation

$$u_{xx} + \frac{1}{x} u_x + \sin u = 0. \tag{0.11}$$

The equation (0.11) represents a particular case of $P\,\overline{\underline{III}}$ equation (see Appendix 3) and is reduced to the latter by the transform

$$u \longrightarrow \exp \frac{iu}{2} .$$

Earlier in the work [28] by V.E.Petrov and the first author an asymptotic parametrization in terms of monodromy data (as $x \longrightarrow +\infty$) has been obtained for a solution to the equation (0.11). It was used in [28] for a calculation of large-time self-similar asymptotic to the Cauchy problem of the Sine-Gordon equation. It is worth mentioning here the possibility of direct calculation of large-time asymptotics of the solutions to evolutionary equations admitting self-similar Painlevé solutions as another application of IDM (see the papers of the first author [29] , [30]). The corresponding procedure is presented in Chapter 12 of the present text , using MKdV equation as an example.

We suggest here our version of IDM concentrating all our attention on two concrete examples which arise most frequently in various physical applications mentioned above. There are the particular case of the $P\text{II}$ equation with $\nu = 0$ and the particular case (0.11) of the $P\text{III}$ equation. From the viewpoint of our analogy between Painlevé transcendents and classical special functions, the equation $P\text{II}$ with $\nu = 0$ corresponds to the Airy function, while equation (0.11) describes a non-linear analogue of the Bessel function with zero symbol.

We outline the most essential results, concerning the two equations, obtained in the main text of the present paper:

a) the complete asymptotic description of real-valued and pure imaginary solutions to equation (0.11) including regular as $x \to +\infty$ solutions as well as those having singularities as $x \to +\infty$, or $x \to 0$. We present the explicit connection formulae linking the asymptotics of solutions (in regular case) and distribution of singularities (in irregular case) with the initial data.

b) the complete description of pure imaginary solutions to $P\text{II}$ equation with $\nu = 0$, i.e. in equivalent terms, the real-valued solutions to the equation

$$u_{xx} - xu + 2u^3 = 0 ,$$

including the explicit calculation of the connection formulae for asymptotics of solutions as $x \to +\infty$ and $x \to -\infty$.

c) the description of the two-parameter set of real-valued $P\,II$ solutions oscillating as $x \to -\infty$ in terms of monodromy data. We extract among them the one-parameter set of solutions, exponentially decreasing as $x \to +\infty$, and present for their asymptotics.*)

The existence of infinite number of poles tending to $+\infty$, is established for the general solution belonging to the real-valued two-parameter set. The asymptotic distribution of poles is given in terms of the leading term parameters of asymptotics at $-\infty$.

Besides the results mentioned above we extend the connection formulae up to a three-parameter set of complex-valued solutions to the equations (0.11) and $P\,II$ with $\nu = 0$. Among them there is a particular solution, first found in [5] by B.M.McCoy et al. and described by some limiting case of our connection formulae. In [5] those formulae were obtained in a rather skillful manner. Note that besides this remarkable work the authors know only one paper [7] where similar formulae were found without any usage of IDM.

The potential possibilities provided by IDM are not exhausted by the results presented in the main text of the paper. In a parallel study accomplished by B.I.Suleimanov by means of IDM the connection formulae have beeen found for a solution to a special type of $P\,V$ equation, arising in statistical physics [7] (see Appendix 1). The case of

*) These formulae appeared for the first time in the paper [31] by M.Ablowitz snd H-Segur who derived them exploiting the analogy mentioned above between PII and MKdV equations. The consistent proof of the connection formulae in a framework of IDM was presented by B-I. Suleimanov in [32] .

$P\,\mathrm{IV}$ equation (see [70]) was studied by A.Kitaev. The complete investigation of real-valued solutions to $P\,\mathrm{II}$ equation with $\nu = 0$ comprising the asymptotic distribution of poles which tend either to $+\infty$, or to $-\infty$, was carried out by A.A.Kapaev.*) In order to give a possibly complete picture of asymptotic results obtained up till now for the set of Painlevé equations, we present the list of connection formulae together with references to the corresponding places in the text where these formulae are proved. For those small number of results known earlier, we give the precise references to papers, where they have appeared first.

For justification of the asymptotic results obtained in the present paper, the following circumstance are of essential importance: the local existence of solutions to Painlevé equations with a prescribed asymptotics may be proved independently of IDM and even appears to be a well-known fact in some particular cases. For example, in papers [33] , [34] the local behaviour of solutions to $P\,\mathrm{II}$ and $P\,\mathrm{III}$ equations in the neighbourhood of infinity has been established. In Chapters 7 - 9 we essentially exploit the results of the work [33] in order to prove our connection formulae, as well as to provide the completeness of asymptotic description of the regular solutions to $P\,\mathrm{II}$ and $P\,\mathrm{III}$ equations. Similarly we make use of the classical results by Paul Painlevé [1] together with modern studies [35] , [36] while justifying in Chapters 10, 11 the asymptotic distributions of poles. To our regret we are unable so far to provide all the proofs within the framework of IDM only. The reason is just the absence of satisfactory solution of the monodromy theory inverse problem for the linear systems associated with Painlevé equations. In general the inverse problems in question are equivalent to certain discontinuous matrix Riemann-Hilbert problems on a set of rays in λ-plane. For example, the Riemann-Hilbert problem for the system (0.4) is defined on the rays $\arg \lambda = \frac{\pi k}{6}$,

*) See Appendix 2.

$k=0,1,\dots,5$. This problem was studied in detail in the work [37] by M.Ablowitz and A.Focas, where they have reduced it to a sequence of three matrix Riemann-Hilbert problems on the real axis. We could not, however, extract from [37] any suitable criteria of solvability of inverse problem, which are necessary for the proof of basic theorems in Chapters 7 - 11. On the other hand, those theorems, being proved independently of the analysis of Riemann-Hilbert problems, provide the solvability criteria in question. In particular the results of Chapter 9 yield that under the conditions on the Stokes matrices $\begin{pmatrix} 1 & 0 \\ p & 1 \end{pmatrix}, \begin{pmatrix} 1 & q \\ 0 & 1 \end{pmatrix}$ (see Chapter 1)

$$p=-\bar{q}\ , \quad x\in\mathbb{R}\ , \quad \nu=0\ ,$$

or

$$p=-q=ia, \quad -1<a<1\ , \quad x\in\mathbb{R}\ , \quad \nu=0$$

the inverse problem for the system (0.4) is uniquely solvable. Similarly the results of Chapter 10 might be interpreted in the same context as an explicit description of those values of deformation parameter x, at which the monodromy theory inverse problem for a given $p,\ q=\bar{p},\ \operatorname{Re} p\neq 0\ ,\ \nu=0$ becomes unsolvable. The application of similar results to inverse problem for the system. associated with $P\text{III}$ equation (0.11) might be derived from Chapters 8, 11.

At the beginning of the Introduction we already mentioned the origin of idea to consider the Painlevé equations as the deformation conditions for the suitable linear systems with rational coefficients. This idea arose first in papers [21] by R.Fuchs and [22] by R. Garnier. However the linear systems used there possess (as a rule) only regular singular points and the deformation parameter x in these systems coinciding with a coordinate of one of the poles. Hence the limiting case $x\to\infty$ does not lead to a significant simplification of the system, and one is unable to accomplish an asymptotic calculations of the monodromy data in this case. Unlike the classical work [21] , [22] , the linear systems, appearing in modern studies [14] , [16] cited above, have irregular singular points, just as the system (0.4). Moreover, the deformation parameter x enters the exponent of essential singularity of solution in the neighbourhood of irregular point (see formulae (1.11'), (1.36) in Chapter 1). This circumstance becomes crucial for the possibility of asymptotic investigation of the monodromy theory direct problem. As a matter of fact the system admits the WKB-solutions which turn to be asymptotical in a double sense: for $x \to \infty$ as well, as for $\lambda \to \infty$. As it usually occurs for similar problems ([38]), this observation is the main technical feature providing explicit calculation of Stokes matrix and derivation at a final stage, of the connection formulae for the Painlevé functions.

In order to give some sort of guidance in the text, we review briefly the structure of the work. The first three chapters play an auxiliary role and do not contain new results. We have included them trying to obtain a self-contained text. In Chapter 1, following the style and notations of the work [14] , we give the definitions of "monodromy data" for the systems of ordinary differential equations with rational coefficients. The detailed description of two concrete systems, denoted as (1.9) and (1.26), which later turn to be associated with P_{II} and P_{III} equations is also presented. In the second Chapter (again following the work [14]) the equations of isomonodromic deformations are derived for a general case of the linear systems. The general formulae obtained here, are applied in Chapter 3 to the systems (1.9) and (1.26). It is proved thus that isomonodromic conditions for (1.9) and (1.26) coincide respectively with the particular case of P_{II} equation $(\nu = 0)$ and P_{III} equation (0.11). First this result has been obtained in [16] by the method described above based on the "self-similar reduction" in $U-V$ pair for MKdV and Sine-Gordon equations.

The fourth Chapter is reserved for the discussions of solvability of the monodromy theory inverse problems for the systems (1.9) and

(1.26). We reproduce there a procedure of the works [16] , [37] which reduces the inverse problem to a certain matrix Riemann-Hilbert problem. The corresponding singular integral equations admit adequate asymptotic analysis as $x \to +\infty$ under the condition

$$p = -q$$

on the scalar values p, q parametrising the conjugation matrices. As a result we construct the two-parameter set of complex-valued solutions to P_{II} equation in terms of monodromy data p, q . These solutions decreasing exponentially as $x \longrightarrow +\infty$. We reveal simultaneously the essence of principal difficulties arising in applications of singular integral equations in the asymptotic analysis of inverse problems.

The asymptotic studies of the monodromy theory direct problems for the systems (1.9), (1.26) are presented in Chapters 5 - 11, which are the main contents of the work. We derive and then prove the asymptotic formulae for various solutions to P_{II} and P_{III} equations. Note that the results and technical approaches used in Chapters 5, 6 and 9 where taken from the papers [30] , [28] , [39] by the first authors and A.A.Kapaev while the results and methods of Chapters 8,10 and 11 are from papers [27],[40]-[42] of the second authors.

The last four Chapters are devoted to an exposition of the most typical applications of modern theory of Painlevé equations. Besides the results obtained in the basic text of the work, we use in these Chapters the results of the following papers:

[43] , [44] by M.Ablowitz and H.Segur, [45] by V.E.Zakharov and S.V. Manakov, [46] by A.B.Shabat - in Chapter 12;

[17] by S.V.Manakov - in Chapter 13 ;

[5] by B.M.McCoy, A.C.Tracy and T.T. Wu - in Chapter 14;

[19] by V.E.Zakharov, E.A.Kuznetsov, S.L.Musher - in Chapter 15.

In Appendices 1 and 2 an application of IDM to the Painlevé V and another type of Painlevé II functions is considered. In Appendix 3 we enumerate (following the work [15]) all six Painlevé equations together with the corresponding "equations in λ ". Finally, in Appendix 4 the list of asymptotic formulae is presented, which contains the connection formulae for the solutions to Painlevé equations. It is provided with references to the corresponding places in the text, where the corresponding formula is derived, together with the reference on the original paper, where it first occurred.

Chapter 1. MONODROMY DATA FOR THE SYSTEMS OF LINEAR ORDINARY DIFFERENTIAL EQUATIONS WITH RATIONAL COEFFICIENTS

Let us fix on $\mathbb{C}P^1$ different points $a_1, \ldots, a_n, a_\infty$ and assign an integer $r_\nu \geq 0$, $\nu \neq \infty$, $r_\infty \geq 1$ to each point a_ν, $\nu = 1, \ldots, n, \infty$. Consider a system of linear ordinary differential equations with rational coefficients:

$$\frac{d\Psi}{d\lambda} = A(\lambda)\Psi , \tag{1.1}$$

$$A(\lambda) = \sum_{\nu=1}^{n} \sum_{k=0}^{r_\nu} \frac{A_{\nu,-k}}{(\lambda - a_\nu)^{k+1}} - \sum_{k=1}^{r_\infty} A_{\infty,-k} \lambda^{k-1} ,$$

where $A_{\nu,-k}$ are $m \times m$ matrices $(m > 1)$ independent of λ. Suppose that a general situation takes place, so that all matrices $A_{\nu,-r_\nu}$ have diagonal jordanian form

$$A_{\nu,-r_\nu} = G^{(\nu)} T^{(\nu)}_{-r_\nu} G^{(\nu)^{-1}} ,$$

$$(T^{(\nu)}_{-r_\nu})_{\alpha\beta} = t^{(\nu)}_{-r_\nu,\alpha} \delta_{\alpha\beta} ,$$

where

$$t^{(\nu)}_{-r_\nu,\alpha} \neq t^{(\nu)}_{-r_\nu,\beta} , \quad \alpha \neq \beta , \quad r_\nu > 0 ,$$

$$t^{(\nu)}_{-r_\nu,\alpha} \neq t^{(\nu)}_{-r_\nu,\beta} \pmod{\mathbb{Z}} , \quad r_\nu = 0 . \tag{1.2}$$

Suppose finally that system (1.1) is calibrated as follows:

$$G^{(\infty)} = I \Rightarrow A_{\infty,-r_\infty} = T^{(\infty)}_{-r_\infty} .$$

In the neighbourhood of each point a_ν, $\nu=1,\dots,n,\infty$ the system (1.1) has a formal solution

$$\Psi_f^{(\nu)}(\lambda) = B^{(\nu)} \cdot \hat{\Psi}^{(\nu)}(\lambda) \exp T^{\nu}(\lambda), \tag{1.3}$$

where

$$T^{\nu}(\lambda) = \operatorname{diag} T^{\nu}(\lambda) = \sum_{k=1}^{r_\nu} T_{-k}^{(\nu)} \frac{z_\nu^{-k}}{(-k)} + T_0^{(\nu)} \ln z_\nu ,$$

$$\hat{\Psi}^{(\nu)}(\lambda) = \sum_{k=0}^{\infty} \psi_k^{(\nu)} z_\nu^k , \qquad \psi_0^{(\nu)} = I ,$$

$$z_\nu = \lambda - a_\nu \ (\nu \neq \infty), \quad z_\infty = \frac{1}{\lambda} , \quad B^{\infty} = I .$$

The matrices $T_0^{(\nu)}$ are called according to [14] the exponents of formal monodromy at the point a_ν, because they indicate the branching of the solution (1.3) at a_ν. It is shown in [14] that under conditions (1.2) all diagonal matrices $T_{-k}^{(\nu)}$ and all coefficients $B^{(\nu)}$ $\psi_k^{(\nu)}$ in the formal series $B^{(\nu)} \cdot \hat{\Psi}^{(\nu)}(\lambda)$ are uniquely determined by a recurrence through the coefficients $A_{\nu,-k}$ of the system (1.1). We are not going to present here this rather cumbersome calculations. Note only that its technique is based on the substitution of the formal series (1.3) into (1.1). The concrete examples would be treated below according to this technique.

Let us pass now to the description of nonformal properties of solutions of the system (1.1). Introduce first some useful notations:

1. $\hat{\mathbb{P}}, \pi : \hat{\mathbb{P}} \to \mathbb{C}P^1/\{a_1,\dots,a_n,\infty\}$ - are respectively the universal covering and the covering transformation for multiconnected domain $\mathbb{C}P^1/\{a_1,\dots,a_n,a_\infty\}$.
2. $\hat{\Omega}_1^{(\nu)},\dots,\hat{\Omega}_{2r_\nu+1}^{(\nu)}$ is the set of sectors on $\hat{\mathbb{P}}$ such that:

a) $\hat{\Omega}_k^{(\nu)} \cap \hat{\Omega}_{k+1}^{(\nu)} \neq \emptyset \quad , \; k=1,\dots,2\tau_\nu ,$

b) $\pi\Big(\bigcup_{k=1}^{2\tau_\nu} \hat{\Omega}_k^{(\nu)}\Big) = V_\nu / \{a_\nu\}$ is some neighbourhood of a_ν,

c) $\Omega_k^{(\nu)} = \pi(\hat{\Omega}_k^{(\nu)})$ for every $\beta,\alpha=1,\dots,m$ contain only one

ray, such that

$$Re\big(t^{(\nu)}_{-\tau_{\nu,\alpha,\beta}} \cdot z_\nu^{-\tau_\nu}\big) = 0 \, , \, t^{(\nu)}_{-\tau_\nu,\alpha,\beta} = t^{(\nu)}_{-\tau_\nu,\alpha} - t^{(\nu)}_{-\tau_\nu,\beta} .$$

For example, taking $\delta > 0$ to be sufficiently small, we may construct the sectors $\hat{\Omega}_{k,\delta}^{(\nu)}$ in the form

$$\pi(\hat{\Omega}_{k,\delta}^{(\nu)}) = \Big\{ \lambda \in V_\nu : \frac{\pi(k-1)}{\tau_\nu} - \delta < \arg z_\nu < \frac{\pi k}{\tau_\nu} \Big\} \; , \; \tau_\nu > 0 .$$

Clearly they satisfy the conditions a) - c). For $\tau_\nu = 0$ we put $\pi(\hat{\Omega}_1^{(\nu)}) = V_\nu / \{a_\nu\}$.

The central result of the theory of linear ordinary systems with rational coefficients ([47] - [49]) may be formulated now as follows:

THEOREM 1.1. Let the conditions (1.2) are satisfied. Then for any point a_ν there may be found a neighbourhood V_ν such that the system (1.1) in every sector $\hat{\Omega}_k^{(\nu)}$ has a solution $\Psi_k^{(\nu)}(\lambda)$, which is holomorfic in λ, matrix-inversible and has $\Psi_f^{(\nu)}(\lambda)$ as its formal asymptotics:

$$\Psi_k^{(\nu)}(\lambda) \cong \Psi_f^{(\nu)}(\lambda), \quad \lambda \to a_\nu \, , \quad \lambda \in \hat{\Omega}_k^{(\nu)} .$$

The asymptotics is uniform in any closed subsector of sector $\hat{\Omega}_k^{(\nu)}$. Every such solution $\Psi_k^{(\nu)}(\lambda)$ may be analitically expanded into all $\hat{\mathbb{P}}$.

Further in the text the solutions $\Psi_k^{(\nu)}(\lambda)$ would be called the canonical solutions of the system (1.1).

REMARK 1.1. The nature of singularity of function $\Psi_k^{(\nu)}(\lambda)$ at the point a_ν considerably depends on the parameter r_ν. In the case when $r_\nu > 0$ the solution $\Psi_k^{(\nu)}$ has essential singularity at a_ν (irregular singular point) and when $r_\nu = 0$ the point a_ν appears to be branching point (regular singular point). In the last case the series for $\hat{\Psi}^{(\nu)}(\lambda)$ converges uniformly and the branching of solution $\Psi_1^{(\nu)}(\lambda)$ at the point a_ν is completely described by corresponding formal monodromy exponent $T_0^{(\nu)}$:

$$\Psi_1^{(\nu)}(\lambda)\Big|_{z_\nu \mapsto e^{2\pi i} z_\nu} = \Psi_1^{(\nu)}(\lambda)\, e^{2\pi i T_0^{(\nu)}} . \tag{1.4}$$

As all the functions $\Psi_k^{(\nu)}(\lambda)$ are the solutions of the same equation (1.1), they are linked with one another by nondegenerate matrix multipliers independent of λ, which fall into two distinct sets:

1. Stokes matrices (multipliers) $S_k^{(\nu)}$, which connect the canonical solutions associated with only one singular point a_ν :

$$\Psi_{k+1}^{(\nu)}(\lambda) = \Psi_k^{(\nu)}(\lambda)\, S_k^{(\nu)} , \quad k = 1, \ldots, 2r_\nu . \tag{1.5}$$

2. Connection matrices $Q^{(\nu)}$, which connect the canonical solutions at infinity with those defined at finite points a_ν, $\nu \neq \infty$:

$$\Psi_1^{(\infty)}(\lambda) = \Psi_1^{(\nu)}(\lambda)\, Q^{(\nu)}, \quad \nu = 1, \ldots, n .$$

Let us note at once the typical triangular structure of Stokes matrices. If we take for σ the unique permutation of m numbers, such that

$$\operatorname{Re} t^{(\nu)}_{-r_\nu, \sigma(1)} \frac{z_\nu^{-r_\nu}}{(-r_\nu)} > \cdots > \operatorname{Re} t^{(\nu)}_{-r_\nu, \sigma(m)} \frac{z^{-r_\nu}}{(-r_\nu)}$$

as $\lambda \in \hat{\Omega}_k^{(\nu)} \cap \hat{\Omega}_{k+1}^{(\nu)}$, then

$$(S_k^{(\nu)})_{\sigma(\alpha)\sigma(\beta)} = \begin{cases} 0, & \alpha < \beta \\ 1, & \alpha = \beta . \end{cases} \tag{1.6}$$

The joint collection of matrices $S_k^{(\nu)}$ and $Q^{(\nu)}$ completed with the collection of $T_0^{(\nu)}$ would be called the monodromy data for equation (1.1).

The monodromy data associated with eg. (1.1) contain the complete information about the global properties of any canonical solution $\Psi^{(\nu)}(\lambda)$, $\Psi_k^{(\infty)}(\lambda)$. In particular, by knowing $S_k^{(\nu)}$, $Q^{(\nu)}$ and $T_0^{(\nu)k}$ we can compute the monodromy group $\mathfrak{M}$ for eg (1.1).

Its generators are provided by M_ν matrices, describing the transformation of solution $\Psi_1^{(\infty)}(\lambda)$ after circling around the point a_ν :

$$\Psi_1^{(\infty)}(\lambda)\Big|_{z_\nu \mapsto e^{2\pi i} z_\nu} = \Psi_1^{(\infty)}(\lambda)\, M_\nu .$$

It is easy to derive the expressions of M_ν using (1.4), (1.5), (1.6) (see [14])

$$M_\nu = [Q^{(\nu)}]^{-1} e^{2\pi i T_0^{(\nu)}} Q^{(\nu)}, \quad r_\nu = 0,$$

$$M_\nu = [Q^{(\nu)}]^{-1} e^{2\pi i T_0^{(\nu)}} [S_1^{(\nu)} S_2^{(\nu)} \dots S_{2r_\nu}^{(\nu)}]^{-1} Q^{(\nu)}, \quad r_\nu > 0. \tag{1.7}$$

The usual cyclic constraint is true

$$M^{(\infty)} M^{(n)} M^{(n-1)} \dots M^{(1)} = I . \tag{1.8}$$

Let us now illustrate the above theory on two concrete examples, which would play the basic role in the main part of the text.

EXAMPLE 1. Consider the particular case of system (1.1)

$$\frac{d\Psi}{d\lambda}=\left\{-(4i\lambda^2+ix+iu^2)\sigma_3-4u\lambda\sigma_2-2w\sigma_1\right\}\Psi\ , \qquad (1.9)$$

where

$$\sigma_1=\begin{pmatrix}0&1\\1&0\end{pmatrix},\ \sigma_2=\begin{pmatrix}0&-i\\i&0\end{pmatrix},\ \sigma_3=\begin{pmatrix}1&0\\0&-1\end{pmatrix}$$

are the Pauli matrices, x is real-valued and u, w are complex-valued parameters.

The system (1.9) is distinguished among other systems (1.1) having the only irregular point of the third order $(r_\infty=3)$ at infinity by the following conditions:

a) the structure of essential singularity is fixed at infinity by the equalities *)

$$\begin{aligned} T_{-1} &= ix\sigma_3\ , \\ T_{-2} &= 0\ , \\ T_{-3} &= 4i\sigma_3\ . \end{aligned} \qquad (1.10)$$

b) there is a specific reduction on matrix $A(\lambda)$:

$$A^T(-\lambda)=A(\lambda). \qquad (1.11)$$

The later condition is derived by substitution into equation

$$\frac{d\Psi}{d\lambda}=[A_{-3}\lambda^2+A_{-2}\lambda+A_{-1}]\Psi$$

a formal series

---

*) There are no other essential singularities exept of infinity, so we shall omit index " ∞ " every where in this example.

$$\Psi_f(\lambda)=\left(I+\psi_1\lambda^{-1}+\psi_2\lambda^{-2}+\ldots\right)\exp\left\{-\frac{4i}{3}\lambda^3\sigma_3-ix\lambda\sigma_3-T_0\ln\lambda\right\}, \tag{1.11'}$$

which yields the constraints

$$A_{-3}=-4i\lambda^2\sigma_3$$

$$A_{-2}=4i[\sigma_3,\psi_1]=4i\begin{pmatrix}0 & u\\ -v & 0\end{pmatrix}, \tag{1.11''}$$

$$u=2(\psi_1)_{12}, \quad v=2(\psi_1)_{21};$$

$$A_{-1}=4i[\sigma_3,\psi_2]-4i[\sigma_3,\psi_1]-ix\sigma_3=$$

$$=-(ix+2iuv)\sigma_3-2\begin{pmatrix}0 & w\\ y & 0\end{pmatrix},$$

$$w=-4i(\psi_2)_{12}+2iu(\psi_1)_{22},$$
$$y=4i(\psi_2)_{21}-2iv(\psi_1)_{11}. \tag{I.I2}$$

According to (1.11) we must putting $v=u$, $y=w$ and so have finally

$$A_{-2}=-4u\sigma_2,$$
$$A_{-1}=-(ix+2iu^2)\sigma_3-2w\sigma. \tag{1.12'}$$

Later on we'll consider the case of real u, w which implies the additional reduction on $A(\lambda)$:

$$\sigma_1\bar{A}(\bar{\lambda})\sigma_1=A(\lambda). \tag{1.13}$$

Let us describe now in detail the collection of monodromy data for equation (1.9). We'll show first that formal monodromy exponent T_0 is equal to zero. It follows immediately from substitution formal series $\Psi_f(\lambda)$ into equation (1.9), which yields

$$T_0 = \operatorname{diag}\left[-4u\sigma_2\psi_2 - 2w\sigma_1\psi_1 - 2iu^2\sigma_3\psi_1\right] =$$

$$= \begin{pmatrix} 4iu(\psi_2)_{21} - wu - 2iu^2(\psi_1)_{11}, & 0 \\ 0, & -4iu(\psi_2)_{12} - wu + 2iu^2(\psi_1)_{22} \end{pmatrix} .$$

Taking into account formulae (1.12) and the basic reduction $u=v$, $w=y$ we get the equality $T_0 = 0$.

The canonical domains $\hat{\Omega}_k$, $k=1,\ldots,7 = 2r_\infty+1$, in the neighbourhood of infinity are introduced as follows (see figure 1):

$$\hat{\Omega}_1 = \{\lambda : -\frac{\pi}{3} < \arg\lambda < \frac{\pi}{3}\} ,$$

$$\hat{\Omega}_2 = \{\lambda : 0 < \arg\lambda < \frac{2\pi}{3}\} ,$$

$$\hat{\Omega}_3 = \{\lambda : \frac{\pi}{3} < \arg\lambda < \pi\} ,$$

$$\hat{\Omega}_4 = \{\lambda : \frac{2\pi}{3} < \arg\lambda < \frac{4\pi}{3}\} ,$$

$$\hat{\Omega}_5 = \{\lambda : \pi < \arg\lambda < \frac{5\pi}{3}\} ,$$

$$\hat{\Omega}_6 = \{\lambda : \frac{4\pi}{3} < \arg\lambda < 2\pi\} ,$$

$$\hat{\Omega}_7 = \{\lambda : \frac{5\pi}{3} < \arg\lambda < \frac{7\pi}{3}\} .$$

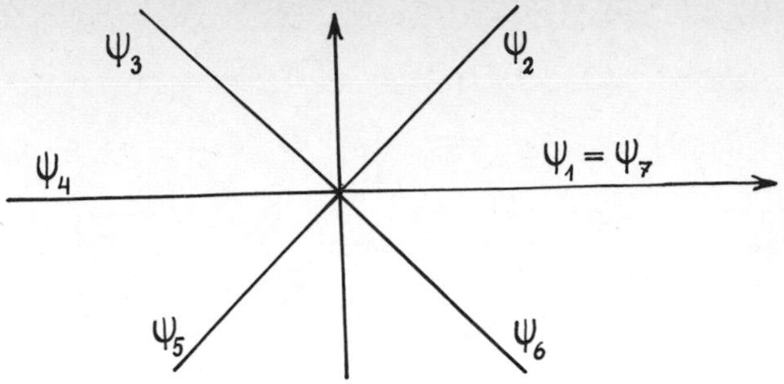

Figure 1.1

According to general theory described above, the canonical solutions $\Psi_k(\lambda)$, $k=1,\dots,7$, are defined as solutions of eq. (1.9) having in corresponding sector $\hat{\Omega}_k$ the asymptotics

$$\Psi_k(\lambda)=(I+\psi_1\lambda^{-1}+\dots)\exp\{-\frac{4i}{3}\lambda^3\sigma_3-ix\lambda\sigma_3\},$$
$$\lambda\to\infty,\ \lambda\in\hat{\Omega}_k. \tag{1.14}$$

The formal monodromy exponent T_0 is zero, so the equality holds

$$\Psi_1(\lambda)=\Psi_7(\lambda) \tag{1.15}$$

and all the matrices $\Psi_k(\lambda)$ appear to be the entire functions of λ. Thus it is possible to consider $\Psi_k(\lambda)$, as been defined directly in the sectors $\Omega_k=\pi(\hat{\Omega}_k)$.

The automorphism $\lambda\mapsto-\lambda$ transforms the sector Ω_k into Ω_{k+3}, so due to equality (1.11) on matrix $A(\lambda)$ the solutions $\Psi(\lambda)$ are transformed as follows:

$$\Psi(\lambda)\mapsto[\Psi^T(-\lambda)]^{-1}. \tag{1.16}$$

The essential singularity (1.14) is invariant under the automorphism (1.16). Bringing together all the facts mentioned above, we obtain the equality on canonical solutions:

$$\Psi_{k+3}(\lambda) = [\Psi_k^T(-\lambda)]^{-1}, \quad k=1,2,3 . \tag{1.17}$$

The complete set of monodromy data for eq. (1.9) is provided by the collection of Stokes matrices

$$S_k = \Psi_k^{-1} \Psi_{k+1}, \quad k=1,\ldots,6 .$$

The constraints (1.17) together with (1.15) show that

$$S_{k+3} = (S_k^T)^{-1}, \quad k=1,2,3 . \tag{1.18}$$

The apriori structure (see the triangularity conditions (1.6)) of matrices S_1, S_2, S_3 has the form

$$S_1 = \begin{pmatrix} 1 & 0 \\ p & 1 \end{pmatrix}, \quad S_2 = \begin{pmatrix} 1 & \tau \\ 0 & 1 \end{pmatrix}, \quad S_3 = \begin{pmatrix} 1 & 0 \\ q & 1 \end{pmatrix}. \tag{1.19}$$

There is the cyclic constraint (1.8) which means in this situation $(T_0 = 0!)$ that the circling around singularity is trivial:

$$S_6 S_5 S_4 S_3 S_2 S_1 = I . \tag{1.20}$$

Applying here the expressions (1.18), (1.19) for S_k it is easy to get from (1.20) a constraint on p, q, τ ([16]):

$$\tau - p - q - \tau p q = 0 .$$

It implies the two destinct cases for its solutions:

a)

$$\tau = \frac{p+q}{1-pq}, \quad pq \neq 1 , \tag{1.21a}$$

b)

$$p = \pm i, \quad q = \mp i . \tag{1.21b}$$

In the case (1.21a) - the general situation case - the monodromy data are parametrized by two complex-valued numbers p, q :

$$S_1 = \begin{pmatrix} 1 & 0 \\ p & 1 \end{pmatrix}, \quad S_2 = \begin{pmatrix} 1 & \frac{p+q}{1-pq} \\ 0 & 1 \end{pmatrix},$$

$$S_3 = \begin{pmatrix} 1 & 0 \\ q & 1 \end{pmatrix}, \quad S_4 = \begin{pmatrix} 1 & -p \\ 0 & 1 \end{pmatrix}, \tag{1.22}$$

$$S_5 = \begin{pmatrix} 1 & 0 \\ \frac{p+q}{pq-1} & 1 \end{pmatrix}, \quad S_6 = \begin{pmatrix} 1 & -q \\ 0 & 1 \end{pmatrix}.$$

Note that system (1.9) has exactly two complex-valued parameters u, w.

The case (1.21b) is a special one. The monodromy data are parametrized by the only one complex-valued number τ and have a specific form:

$$S_1 = \begin{pmatrix} 1 & 0 \\ \pm i & 1 \end{pmatrix}, \quad S_2 = \begin{pmatrix} 1 & \tau \\ 0 & 1 \end{pmatrix},$$

$$S_3 = \begin{pmatrix} 1 & 0 \\ \mp i & 1 \end{pmatrix}, \quad S_4 = \begin{pmatrix} 1 & \mp i \\ 0 & 1 \end{pmatrix}, \tag{1.23}$$

$$S_5 = \begin{pmatrix} 1 & 0 \\ -\tau & 1 \end{pmatrix}, \quad S_6 = \begin{pmatrix} 1 & \pm i \\ 0 & 1 \end{pmatrix}.$$

At the end of our consideration of example 1 let us discuss the real-valued reduction on monodromy data. As we have mentioned above the equalities $u=\bar{u}$, $w=\bar{w}$ are equivalent to the constraint (1.13) on matrix $A(\lambda)$, which in its turn implies the existence of additional automorphism on the set of solutions of eq. (1.9)

$$\Psi(\lambda) \mapsto \sigma_1 \Psi(\bar{\lambda}) \sigma_1 .$$

It produces thus additional relations between the canonical solutions:

$$\Psi_6(\lambda) = \sigma_1 \overline{\Psi}_2(\bar{\lambda})\sigma_1 ,$$

$$\Psi_5(\lambda) = \sigma_1 \overline{\Psi}_3(\bar{\lambda})\sigma_1 ,$$

$$\Psi_4(\lambda) = \sigma_1 \overline{\Psi}_1(\bar{\lambda})\sigma_1 .$$

These equalities imply that

$$S_6 = \Psi_6^{-1}(\lambda)\Psi_7(\lambda) = \Psi_6^{-1}(\lambda)\Psi_1(\lambda) =$$

$$= \sigma_1 \overline{\Psi}_2^{-1}(\bar{\lambda})\overline{\Psi}_1(\bar{\lambda})\sigma_1 = \sigma_1 \overline{S}_1^{-1}\sigma_1 , \tag{1.24}$$

$$S_3 = \sigma_1 \overline{S}_1^{T} \sigma_1 \Leftrightarrow q = \bar{p},$$

$$S_5 = \Psi_5^{-1}\Psi_6 = \sigma_1 \overline{\Psi}_3^{-1}(\bar{\lambda})\overline{\Psi}_2(\bar{\lambda})\sigma_1 =$$

$$= \sigma_1 \overline{S}_2^{-1}\sigma_1 , \tag{1.25}$$

$$S_2 = \sigma_1 \overline{S}_2^{T} \sigma_1 \Leftrightarrow r = \bar{r} .$$

Thus the general real-valued case is described by equalities (1.22), where $p = \bar{q}$, $|p|^2 \neq 1$. The special case leads to the reduction (1.23), where $r = \bar{r}$.

REMARK 1.2. Alongside with the real-valued reduction (1.13) in eq. (1.9) we would be interested in the case of purely imaginary u and w. It is easy to prove that corresponding monodromy data are fully described by eqs. (1.22) where $p = -\bar{q}$.

EXAMPLE 2. We take here the system (1.1) in the form

$$\frac{d\Psi}{d\lambda} =$$

$$= \left\{ -i\frac{x^2}{16}\sigma_3 - \frac{1}{2\lambda}\cdot\frac{ixw}{2}\sigma_1 + \frac{i}{\lambda^2}\cos u\,\sigma_3 - \frac{i}{\lambda^2}\sin u\,\sigma_2 \right\}\Psi , \tag{1.26}$$

where σ_k are the same as in example 1, x is real-valued and u, w - complex-valued parameters.

Again we point out the conditions distinguishing eq. (1.26) from 2×2 matrix equations with two irregular singular points of the first order $\lambda = 0$, $\lambda = \infty$:

a) the structure of essential singularities of solution $\Psi(\lambda)$ is fixed by the equalities

$$T_{-1}^{(\infty)} = \frac{ix^2}{16}\sigma_3 , \quad T_{-1}^{(0)} = i\sigma_3 , \tag{1.27}$$

b) the matrix $A(\lambda)$ satisfies the reduction constraint

$$A(\lambda) = -\sigma_1 A(-\lambda)\sigma_1 . \tag{1.28}$$

It may be proved by the similar calculations as being performed above in example 1. The substitution into equation

$$\frac{d\Psi}{d\lambda} = \left[A_{\infty,-1} + A_{0,0}\frac{1}{\lambda} + A_{0,-1}\frac{1}{\lambda^2}\right]\Psi$$

the formal series

$$\Psi_f^{(\infty)}(\lambda) = \left(I + \psi_1^{(\infty)} \cdot \lambda^{-1} + \ldots\right) exp\left\{-\frac{ix^2}{16}\sigma_3\lambda\right\}$$

provides the equalities

$$A_{\infty,-1} = -i\frac{x^2}{16}\sigma_3 ,$$

$$A_{0,0} = -\frac{ix^2}{16}\left[\psi_1, \sigma_3\right] = \begin{pmatrix} 0 & -\frac{ix}{4}w \\ -\frac{ix}{4}v & 0 \end{pmatrix} , \tag{1.29}$$

where

$$w = -\frac{x}{2}\left(\psi_1^{(\infty)}\right)_{12} ,$$

$$v = \frac{x}{2}\left(\psi_1^{(\infty)}\right)_{21} .$$

The reduction (1.28) imposed on $A(\lambda)$ is equivalent thus to $v = w$ and, consequently

$$A_{0,0} = -\frac{ixw}{4}\sigma_1 .$$

Simultaneously it is necessary to calculate the formal solution at $\lambda = 0$:

$$\Psi_f^{(0)}(\lambda) = B^{(0)}(I + \ldots)\exp\{-\frac{i\sigma_3}{\lambda}\} ,$$

which leads to equations

$$A_{0,-1} = i B^{(0)} \sigma_3 [B^{(0)}]^{-1} ,$$

providing

$$\det A_{0,-1} = 1 , \qquad \mathrm{Sp}\, A_{0,-1} = 0 . \tag{1.30}$$

Taking into account the reduction (1.28), there appears the third condition on $A_{0,1}$:

$$\sigma_1 A_{0,-1} \sigma_1 = -A_{0,-1} . \tag{1.31}$$

Bringing together (1.30), (1.31) it is possible to write out $A_{0,-1}$ in the form

$$A_{0,-1} = \begin{pmatrix} i\cos u & -\sin u \\ \sin u & -i\cos u \end{pmatrix}, \quad u \in \mathbb{C} . \tag{1.32}$$

Again as above in the case of eq. (1.9) we should be specially interested in the real-valued reduction of eq. (1.26) $u=\bar{u}$, $w=\bar{w}$. Here this reduction is satisfied when

$$A(\lambda) = -\overset{*}{A}(\bar{\lambda}) \tag{1.33}$$

in addition to (1.28).

Before passing to a description of monodromy data for eq. (1.26) it is worth mentioning two points:

1. Just as in the previous example the formal monodromy exponent is equal to zero at both singular points. We are going to omit the proof as it is rather similar to the proof of the equality $T_0=0$ for eq. (1.9) but seems to be more cumbersome.

2. There are two arbitrary parameters C_1, C_2 in the formal solution $\psi_f^{(0)}$ which is fixed by the matrix $B_\bullet^{(0)}$:

$$B_\bullet^{(0)} = \begin{pmatrix} \cos\frac{u}{2} & -i\sin\frac{u}{2} \\ -i\sin\frac{u}{2} & \cos\frac{u}{2} \end{pmatrix} \begin{pmatrix} C_1 & 0 \\ 0 & C_2 \end{pmatrix}, \quad C_1, C_2 \in \mathbb{C}.$$

We shall exclude that arbitrariness putting $C_1 = C_2 = 1$. Note that under this choice the matrix $B^{(0)}$ would satisfy the identity

$$\sigma_1 B^{(0)} \sigma_1 = B^{(0)}, \tag{1.34}$$

and, assuming u, w been real-valued, the additional equality is true *)

$$\left[B^{(0)*}\right]^{-1} = B^{(0)}. \tag{1.35}$$

Let us get down to the monodromy data. Due to the equality $T_0^{\infty,0} = 0$ the complete set of monodromy data for eq. (1.26) consists

*) The constraint (1.34), (1.35) fix $B^{(0)}$ just uniquely leaving only one scalar multiplier of the form $e^{i\alpha}$ been arbitrary.

of four Stokes matrices $S^{(\infty)}_{1,2}$, $S^{(0)}_{1,2}$ and one connection matrix $Q \equiv Q^{(0)}$. The canonical sectors $\hat{\Omega}^{(\infty,0)}_{1,2,3}$ in the neighbourhoods of origin and infinity are defined as follows:

$$\hat{\Omega}^{(\infty,0)}_{1} = \{ \lambda : -\pi < \arg\lambda < \pi, \ 0 \neq |\lambda| \gtrless \rho > 0 \},$$

$$\hat{\Omega}^{(\infty,0)}_{2} = \{ \lambda : 0 < \arg\lambda < 2\pi, \ 0 \neq |\lambda| \gtrless \rho > 0 \},$$

$$\hat{\Omega}^{(\infty,0)}_{3} = \{ \lambda : \pi < \arg\lambda < 3\pi, \ 0 \neq |\lambda| \gtrless \rho > 0 \}.$$

The projections $\Omega_k = \pi(\hat{\Omega}_k)$ of the sectors on the λ plane are shown at the figure 2:

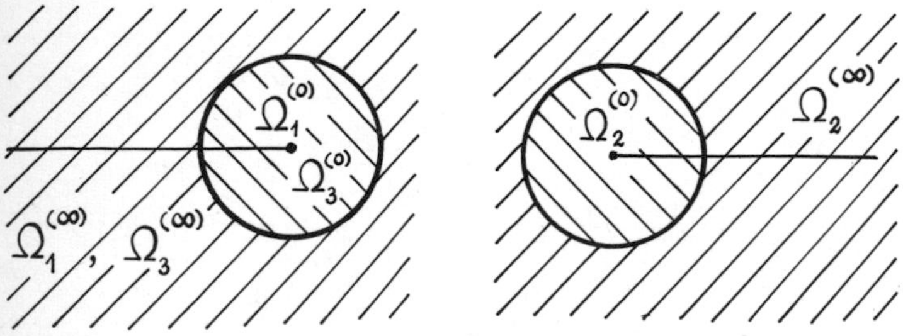

Figure 1.2

The canonical solutions $\Psi^{(\infty,0)}_{k}(\lambda)$ are fixed at the corresponding sectors $\hat{\Omega}^{(\infty,0)}_{k}$ by their asymptotics

$$\Psi^{(\infty)}_{k} = (I + \psi^{(\infty)}_{1}\lambda^{-1} + \ldots)\exp\{-\frac{ix^2}{16}\sigma_3\lambda\}, \quad \lambda \in \hat{\Omega}^{(\infty)}_{k},$$
$$\lambda \to \infty,$$

$$\Psi^{(0)}_{k} = B^{(0)}(I + \lambda\psi^{(0)}_{1} + \ldots)\exp\{-\frac{i\sigma_3}{\lambda}\}, \quad \lambda \in \hat{\Omega}^{(0)}_{k}, \tag{1.36}$$
$$\lambda \to 0,$$

$$B^{(0)} = \begin{pmatrix} \cos\frac{u}{2} & -i\sin\frac{u}{2} \\ -i\sin\frac{u}{2} & \cos\frac{u}{2} \end{pmatrix}, \quad k=1,2,3.$$

The Stokes matrices are as usual

$$S_k^{(\infty,0)} = \Psi_k^{-1}\Psi_{k+1}, \quad k=1,2, \tag{1.37}$$

and the connection matrix Q satisfies the equation

$$\Psi_1^{(\infty)} = \Psi_1^{(0)} Q. \tag{1.38}$$

Despite of the fact that here, as in the previous example, the formal monodromy exponents are trivial the solutions Ψ_k are by no means the entire functions of λ. Their correct global prolongation is possible only on the universal covering $\hat{\mathbb{P}}$. In particular, the equality analogous to eq. (1.15) for eq. (1.26) has to be presented as follows:

$$\Psi_3^{(\infty,0)}(\lambda) = \Psi_1^{(\infty,0)}(\lambda e^{-2\pi i}), \tag{1.39}$$

where $\pi < \arg\lambda < 3\pi$.

From eq. (1.39) and definition (1.37) of the Stokes matrices we extract relations between the functions $\Psi_1^{(\infty,0)}(\lambda)$ on different sheets of $\hat{\mathbb{P}}$:

$$\Psi_1^{(\infty,0)}(\lambda) S_1^{(\infty,0)} S_2^{(\infty,0)} = \Psi_1^{(\infty,0)}(\lambda e^{-2\pi i}), \quad \lambda \in \hat{\Omega}_3 .$$

In other words the monodromy group here is nontrivial and it has two generators

$$M^{(\infty)} = [S_2^{(\infty)}]^{-1} [S_1^{(\infty)}]^{-1},$$

$$M^{(0)} = Q^{-1} \cdot [S_1^{(0)} S_2^{(0)}]^{-1} Q .$$

In contradiction with the general theory we have chosen here the inverse direction of passing around the point $\lambda = \infty$. So the cyclic relation (1.8) is rewritten in this case as

$$M^{(\infty)} = M^{(0)} , \tag{1.40}$$

and it expresses the equivalence of two ways of passing around $\lambda = 0$ and $\lambda = \infty$.

It is convenient for future considerations to introduce alongside with solutions $\Psi_k^{(\infty,0)}$ the auxiliary pair $\Psi_\pm(\lambda)$ fixed by conditions

$$\begin{aligned} &\Psi_\pm(\lambda) = (I + \psi_1^{\pm(\infty)} \lambda^{-1} + \ldots) \exp\{-\frac{ix^2}{16} \sigma_3 \lambda\} , \\ &\qquad \lambda \to \infty , \ \lambda \in \hat{\Omega}_\pm , \\ &\Psi_\pm(\lambda) = B_\pm (I + \psi^{\pm(0)} \lambda + \ldots) \exp\{-\frac{i\sigma_3}{\lambda}\} , \\ &\qquad \lambda \to 0 , \ \lambda \in \hat{\Omega}_\pm , \end{aligned} \tag{1.41}$$

where (see figure 3)

$$\hat{\Omega}_+ = \{ \lambda : 0 < \arg \lambda < \pi , \ \lambda \neq 0 \} ,$$

$$\hat{\Omega}_- = \{ \lambda : -\pi < \arg \lambda < 0 , \ \lambda \neq 0 \}$$

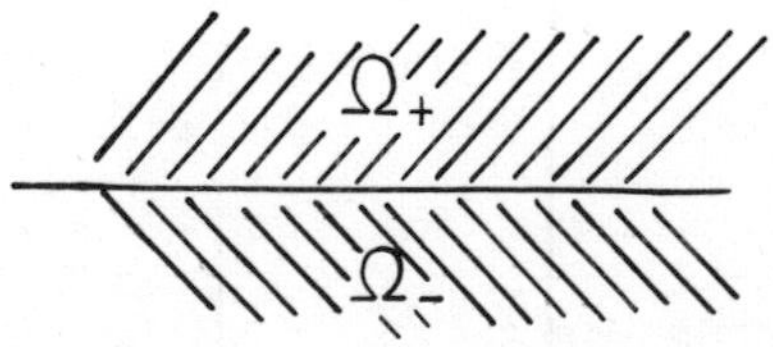

Figure 1.3

Note that the solutions $\Psi_{\pm}(\lambda)$ are defined uniquely by their asymptotics (1.41) in spite of absense of rays with $Im\lambda = 0$ in the domains $\Omega_{\pm}$. In fact if there is another one solution $\widetilde{\Psi}_{+}(\lambda)$ with the same asymptotic behaviour as for $\Psi_{+}(\lambda)$, then for the matrix

$$S = \widetilde{\Psi}_{+}^{-1}\Psi_{+}$$

we derive immediately from the first formula (1.41) that it has upper-triangular structure, whereas from the second formula (1.41) it appears to be lower- triangular. It yields then $S = I$. Note also that $B_{\pm}$ matrices do not in general coinside with $B^{(0)}$:

$$B^{(0)} = B_{\pm}\Lambda_{\pm} ,$$

where $\Lambda_{\pm} = diag\{\lambda_1^{\pm}, \lambda_2^{\pm}\}$. We can make only one apriori statement about the matrices $\Lambda_{\pm}$:

$$det\,\Lambda_{\pm} = 1, \quad \lambda_1^{\pm}\lambda^{\pm} = 1 . \tag{1.42}$$

Let us define the auxiliary Stokes matrices $S_{\pm}^{(\infty,0)}$ linking solutions $\Psi_{\pm}$ with that of $\Psi_1^{(\infty,0)}$:

$$\Psi_{\pm}(\lambda) = \Psi_1^{(\infty)}(\lambda)S_{\pm}^{(\infty)}, \quad \lambda \in \hat{\Omega}_{\pm} \cap \hat{\Omega}_1^{(\infty)},$$

$$\Psi_{\pm}(\lambda)\Lambda_{\pm} = \Psi_1^{(0)}(\lambda)S_{\pm}^{(0)}, \quad \lambda \in \hat{\Omega}_{\pm} \cap \hat{\Omega}_1^{(0)} . \tag{1.43}$$

We'll show now that $S_{-}^{(\infty)}$, $S_{-}^{(0)}$ matrices together with all $S_k^{(\infty,0)}$ Stokes matrices and connection matrix Q are explicitly calculated through the matrices $S_{+}^{(\infty)}$, $S_{+}^{(0)}$. The most simple matter is the connection matrix Q :

$$Q = [\Psi_1^{(0)}]^{-1}\Psi_1^{(\infty)} = [\Psi_{+}\Lambda_{+}S_{+}^{(0)-1}]^{-1}\Psi_{+}(\lambda)S_{+}^{(\infty)^{-1}}$$

so we have

$$Q = S_+^{(0)} \Lambda_+^{-1} [S_+^{(\infty)}]^{-1} . \tag{1.44}$$

In order to obtain constraints for the rest of our matrices it is necessary to use the reduction (1.28) which has the form

$$\Psi(\lambda) \longmapsto \sigma_1 \Psi(-\lambda) \sigma_1 . \tag{1.45}$$

As we have mentioned above the solutions $\Psi_k^{(\infty,0)}(\lambda)$ and $\Psi_{\pm}(\lambda)$ are correctly defined only in the universal covering $\hat{\mathbb{P}}$. That is why the automorphism $\lambda \longmapsto -\lambda$ has to be expanded on $\hat{\mathbb{P}}$ and its action on the solutions $\Psi_k^{(\infty,0)}$ has to be treated with an accuracy. Taking into account the fact of invariance of essential singularity in the formal solutions $\Psi_f^{(\infty)}$ we obtain the reductions

$$\Psi_-(\lambda) = \sigma_1 \Psi_+(e^{i\pi}\lambda)\sigma_1 , \quad -\pi < \arg \lambda < 0 , \tag{1.46}$$

$$\Psi_2^{(\infty)}(\lambda) = \sigma_1 \Psi_1^{(\infty)}(e^{-i\pi}\lambda)\sigma_1 , \quad 0 < \arg \lambda < 2\pi . \tag{1.47}$$

The formulae (1.46), (1.47) imply

$$\Psi_1^{(\infty)} S_+^{(\infty)} = \Psi_+(\lambda) = \sigma_1 \Psi_-(e^{-i\pi}\lambda)\sigma_1 = \sigma_1 \Psi_1^{(\infty)}(e^{-i\pi}\lambda) S_-^{(\infty)} \sigma_1 =$$

$$= \Psi_2^{(\infty)}(\lambda) \sigma_1 S_-^{(\infty)} \sigma_1 , \qquad 0 < \arg \lambda < \pi .$$

That yields immediately

$$S_1^{(\infty)} = S_+^{(\infty)} \sigma_1 [S_-^{(\infty)}]^{-1} \sigma_1 . \tag{1.48}$$

Assuming then $\pi < \arg \lambda < 2\pi$ in (1.46) we get

$$\Psi_1^{(\infty)}(e^{-2\pi i}\lambda)\,S_-^{(\infty)} = \Psi_-(e^{-2\pi i}\lambda) = \sigma_1\Psi_+(\lambda e^{-\pi i})\sigma_1 =$$

$$= \sigma_1\Psi_1^{(\infty)}(\lambda e^{-\pi i})\,S_+^{(\infty)}\sigma_1 = \Psi_2^{(\infty)}(\lambda)\,\sigma_1\,S_+^{(\infty)}\sigma_1 \,. \tag{1.49}$$

This formula together with (1.39) leads to equality

$$\Psi_3^{(\infty)}(\lambda) = \Psi_2^{(\infty)}(\lambda)\sigma_1\,S_+^{(\infty)}\sigma_1\,[S_-^{(\infty)}]^{-1} \,, \tag{1.50}$$

where $\pi < \arg\lambda < 2\pi$.

In other words,

$$S_2^{(\infty)} = \sigma_1\,S_+^{(\infty)}\sigma_1\,[S_-^{(\infty)}]^{-1} \,. \tag{1.51}$$

The formulae (1.48), (1.51) solve the problem of expression of $S_{1,2}^{(\infty)}$ matrices through $S_\pm^{(\infty)}$. The corresponding constraints between the Stokes matrices near the origin look very similar:

$$\begin{aligned} S_1^{(0)} &= S_+^{(0)}\sigma_1\,[S_-^{(0)}]^{-1}\sigma_1 \,, \\ S_2^{(0)} &= \sigma_1\,S_+^{(0)}\sigma_1\,[S_-^{(0)}]^{-1} \,. \end{aligned} \tag{1.52}$$

Their proof just repeats our calculations above which have lead us to eqs. (1.48), (1.51). It uses however the invariance of $B^{(0)}$ matrix (see (1.34)) under the automorphism and the relation between $\Lambda_\pm$

$$\Lambda_- = \sigma_1\Lambda_+\sigma_1 \,, \tag{1.53}$$

which follows from (1.46).

The formula (1.43) may be treated as a way of prolongation of functions $\Psi_\pm(\lambda)$ onto the half-line $\lambda > 0$. Thus the following equations occur

$$\Psi_-(\lambda) = \Psi_+(\lambda)G_+^{(\infty)} \,, \quad G_+^{(\infty)} = [S_+^{(\infty)}]^{-1}\,S_-^{(\infty)} \,, \quad \lambda > 0$$

$$\Psi_-(\lambda)=\Psi_+(\lambda)G_+^{(0)}\ ,\quad G_+^{(0)}=\Lambda_+\ [S_+^{(0)}]^{-1}S_-^{(0)}\Lambda_-^{-1},\quad \lambda>0\ , \tag{1.54}$$

which lead to equality

$$G_+^{(0)}=G_+^{(\infty)}=G_+\ . \tag{1.55}$$

The apriori structure (see (1.6)) of $S_\pm^{(0,\infty)}$ matrices is

$$S_+^{(\infty)}=\begin{pmatrix}1 & 0\\ q & 1\end{pmatrix},\quad S_-^{(\infty)}=\begin{pmatrix}1 & f\\ 0 & 1\end{pmatrix},$$

$$S_+^{(0)}=\begin{pmatrix}1 & p\\ 0 & 1\end{pmatrix},\quad S_-^{(0)}=\begin{pmatrix}1 & 0\\ g & 1\end{pmatrix}. \tag{1.56}$$

Substituting them into (1.55) and making use of (1.42), we get

$$f=-p\ ,\quad g=-q\ ,\quad pq\neq-1\ ,$$

$$\lambda_1^+=\frac{1}{\sqrt{1+pq}}\ ,\quad \lambda_2^-=\sqrt{1+pq}\ . \tag{1.57}$$

Clearly the conditions (1.57) provide the cyclic relation (1.40) to be satisfied identically for any $p\ ,q$.

Summing up the results of the example 2 been discussed, we write down final expressions for the monodromy data of eq. (1.26)

$$S_1^{(\infty)}=\begin{pmatrix}1 & 0\\ p+q & 1\end{pmatrix},\quad S_2^{(\infty)}=\begin{pmatrix}1 & p+q\\ 0 & 1\end{pmatrix},$$

$$S_1^{(0)}=\begin{pmatrix}1 & p+q\\ 0 & 1\end{pmatrix},\quad S_2^{(0)}=\begin{pmatrix}1 & 0\\ p+q & 1\end{pmatrix}, \tag{1.58}$$

$$Q=(1+pq)^{-\frac{1}{2}}\begin{pmatrix}1 & p\\ -q & 1\end{pmatrix},$$

where $p,\ q\in\mathbb{C},\ pq\neq -1$, are non-diagonal parameters of $S_+^{(\infty)},\ S_+^{(0)}$ matrices

$$S_+^{(\infty)}=\begin{pmatrix}1 & 0\\ q & 1\end{pmatrix},\qquad S_+^{(0)}=\begin{pmatrix}1 & p\\ 0 & 1\end{pmatrix}.$$

The auxiliary Stokes matrices $S_-^{(0,\infty)}$ have the form

$$S_-^{(\infty)}=\begin{pmatrix}1 & -p\\ 0 & 1\end{pmatrix},\qquad S_-^{(0)}=\begin{pmatrix}1 & 0\\ -q & 1\end{pmatrix}.$$

After imposition of real-valued reduction (1.33) the specific relations occur

$$\Psi_-(\lambda)=[\Psi_+^*(\bar{\lambda})]^{-1},$$

$$\Psi_1^{(\infty)}=[\Psi_1^*(\bar{\lambda})]^{-1},$$

$$-\pi<\arg\lambda<0,\qquad \arg\bar{\lambda}=-\arg\lambda .$$

It follows then for Stokes matrices that

$$S_-^{(\infty)}=[S_+^{(\infty)*}]^{-1}\Leftrightarrow p=\bar{q}\ . \tag{1.59}$$

REMARK 1.3. Similar to the example 1 the purely imaginary reduction in eq. (1.26) corresponds the monodromy data restriction

$$p=-\bar{q}\ .$$

Chapter 2. ISOMONODROMIC DEFORMATIONS OF SYSTEMS OF LINEAR ORDINARY DIFFERENTIAL EQUATIONS WITH RATIONAL COEFFICIENTS

Suppose that coefficients of eq. (1.1) and positions of its singular points a_ν, $\nu=1,\dots,n$, depend on a certain additional parameter t (scalar or vector):

$$A(\lambda)=A(\lambda;t)\ .$$

Moreover, this dependence is chosen in such a way that all the monodromy data $S_k^{(\nu)}$, $Q^{(\nu)}$ and $T_0^{(\nu)}$ remain constant. Then the global solution $\Psi(\lambda)=\Psi(\lambda;t)$ of eq. (1.1) produces a differential form

$$\Omega(\lambda)=d\Psi\cdot\Psi^{-1}\ ,$$

where

$$d\Psi=\sum_j \frac{\partial\Psi(\lambda;t)}{\partial t_j}dt_j\quad , \tag{2.1}$$

and $\Omega(\lambda)$ is matrix-valued rational in λ form. Its poles lie at the points $a_1,\dots,a_n$, and its decomposition with respect of the leading terms

$$\Omega(\lambda)=\Omega^{(\infty)}+\sum_{\nu=1}^{n}\Omega^{(\nu)}$$

is obviously given by formulas

$$\Omega^{(\infty)}=\sum_{K=0}^{r_\infty}\lambda^K\Omega_K^{(\infty)}\equiv\hat{\Psi}^{(\infty)}(\lambda)\cdot dT^{(\infty)}(\lambda)\hat{\Psi}^{(\infty)^{-1}}(\lambda)(\mathrm{mod}\,\lambda^{-1})$$

$$\Omega^{(\nu)}=\sum_{K=1}^{r_\nu+1}\frac{\Omega_K^{(\nu)}}{(\lambda-a_\nu)^K}\equiv \tag{2.2}$$

$$\equiv B^{(\nu)}\hat{\Psi}^{(\nu)}(\lambda)\,dT^{(\nu)}(\lambda)\,\hat{\Psi}^{(\nu)^{-1}}(\lambda)\,B^{(\nu)^{-1}}(mod\,(\lambda-a_\nu)^{0}\ ),$$

where

$$dT^{(\infty)}(\lambda)=\sum_{K=1}^{r_\infty}\frac{\lambda^{K}}{-K}\,dT^{\infty}_{-K}(t),$$

$$dT^{(\nu)}(\lambda)=-\sum_{K=0}^{r_\nu}\frac{T^{(\nu)}_{-K}}{(\lambda-a_\nu)^{K+1}}\,da_\nu(t)-\sum_{K=1}^{r_\nu}\frac{1}{K(\lambda-a_\nu)^{K}}\,dT^{(\nu)}_{-K}(t).$$

It was metioned above (in Chapter 1) that all matrices $T^{(\nu)}_{-K}$ and all coefficients $B^{(\nu)}\ \psi^{(\nu)}_{K}$ in the series for $B^{(\nu)}\psi^{(\nu)}(\lambda)$ are constructed uniquely and explicitly through the coefficients of eq.(1.1). That is why the form $\Omega(\lambda)$ is also determined uniquely and explicitely through the given $A(\lambda)$ matrix. The constraint (2.1) now may be treated as a fact that $\Psi(\lambda;t)$ satisfies an auxiliary linear system in a parameter t :

$$d\Psi=\Omega(\lambda)\Psi\ . \tag{2.3}$$

The condition of compatibility of eq. (2.3) with the initial eq. (1.1) leads to a constraint

$$dA=\frac{\partial\Omega}{\partial\lambda}+[\Omega,A]\ . \tag{2.4}$$

The latter produces a system of nonlinear equations on matrix coefficients $A_{\nu,-K}$ entering a series expansion of $A(\lambda)$ matrix. As it was shown just above the equation (2.4) provides a necessary condition for initial eq. (1.1) been isomonodromic with respect to the dependence of t . It was proved in [14] that eq. (2.4) is also a sufficient condition of isomonodromy. Moreover, it is shown in [14] that eq.

(2.4) is integrable in the sense of Frobenius and the maximum independent collection of deformation parameters t may always be chosen as a set of singularity point $a_1, \dots, a_n$ together with matrix elements of diagonal matrices $T^{(\nu)}_{-k}$, $k = 1, \dots, r_\nu > 0$.

REMARK 2.1. The particular case of absense of irregular singularities, i.e. when

$$A(\lambda) = \sum_{\nu=1}^{n} \frac{A_\nu}{(\lambda - a_\nu)} ,$$

the equation of isomonodromic deformations (2.4) was considered at the works of Schlesinger [50] . The differential form is

$$\Omega(\lambda) = \sum_{\nu=1}^{n} \Omega^{(\nu)}(\lambda) = -\sum_{\nu=1}^{n} \frac{A_\nu}{\lambda - a_\nu}\, da_\nu$$

and eq. (2.4) is equivalent to the following nonlinear system on A_ν matrices

$$dA_\nu = \sum_{\substack{\mu \neq \nu \\ \mu = 1}} [A_\mu, A_\nu] \frac{da_\mu - da_\nu}{a_\mu - a_\nu}, \quad \nu = 1, \dots, n . \tag{2.5}$$

The system (2.5) is called the system of Schlesinger equations.

REMARK 2.2. For some applications it is important to know how calibration matrices $B^{(\nu)}$ change under isomonodromic deformations. Corresponding equations may be derived directly from the equality

$$d\Psi \cdot \Psi^{-1} = \Omega(\lambda) .$$

Namely, expanding both parts of this equation in the neighboudhood of the point a_ν and scaling terms of zero order in $(\lambda - a_\nu)$, we find

$$dB^{(\nu)} \cdot [B^{(\nu)}]^{-1} = \theta^{(\nu)} , \tag{2.6}$$

$$\overset{(\nu)}{\theta} \equiv \overset{(\infty)}{\Omega}(a_\nu) + \sum_{\substack{\mu=1 \\ \mu\neq\nu}}^{n} \overset{(\mu)}{\Omega}(a_\nu) + \overset{(\nu)}{B}\overset{(\nu)}{\psi_1}[\overset{(\nu)}{B}]^{-1} da_\nu - \overset{\nu}{\omega_0},$$

where $\overset{\nu}{\omega_0}$ is a coefficient at the zero-order term of $(\lambda - a_\nu)$ in the series for the form $\overset{(\nu)}{B}\overset{\wedge(\nu)}{\Psi}(\lambda)\cdot d\overset{(\nu)}{T}(\lambda)[\overset{\wedge(\nu)}{\Psi}(\lambda)]^{-1}[\overset{(\nu)}{B}]^{-1}$. All the forms $\overset{(\nu)}{\theta}$ as well as the form $\Omega(\lambda)$ are expressed explicity through $A_{\nu,-k}$. For example, if there are no irregular singular points, the expression is

$$\overset{\nu}{\omega_0} = \overset{(\nu)}{B}[\overset{(\nu)}{T_0}, \overset{(\nu)}{\psi_1}](\overset{(\nu)}{B})^{-1} da_\nu .$$

On the other hand, in this case the initial equation (1.1) implies the following identity

$$\overset{(\nu)}{B}(\overset{(\nu)}{\psi_1} - [\overset{(\nu)}{T_0}, \overset{(\nu)}{\psi_1}])(\overset{(\nu)}{B})^{-1} = \sum_{\substack{\mu=1 \\ \mu\neq\nu}}^{n} \frac{A_\mu}{a_\nu - a_\mu} .$$

Consequently in the situation of Schlesinger equations (2.5), we have

$$\overset{(\nu)}{\theta} = \sum_{\substack{\mu=1 \\ \mu\neq\nu}}^{n} A_\mu \frac{da_\mu - da_\nu}{a_\mu - a_\nu} .$$

REMARK 2.3. In the case of vector-valued parameter $t=(t_1,\dots,t_m)$ alongside with the "basic" compatibility condition (2.4) there appears an additional identity

$$d\Omega(\lambda) = \Omega(\lambda)\wedge\Omega(\lambda) \tag{2.7}$$

which is a Frobenius-type compatibility condition for eq. (2.3). It is shown in [14] that eq. (2.7) is a corollary of the "basic" compatibility condition (2.4).

Writing down the form $\Omega(\lambda)$ as

$$\Omega(\lambda) = \sum_{j=1}^{m} U_j(\lambda)\, dt_j$$

it is possible to rewrite eq. (2.7) as follows

$$\frac{\partial U_j(\lambda)}{\partial t_k} - \frac{\partial U_k(\lambda)}{\partial t_j} = [U_k, U_j](\lambda). \tag{2.8}$$

This is a system of nonlinear partial differential equations on matrix coefficients of rational functions $U_j(\lambda)$. Moreover, it is exactly the system integrable by the inverse scattering problem (see [51]). Thus the equations of isomonodromic deformations produce a new class of solutions of exactly integrable systems - the class of isomonodromic solutions. Their detailed discussion and determination of their place among other solutions of integrable systems is contained in [30] .

Chapter 3. ISOMONODROMIC DEFORMATIONS OF SYSTEMS (1.9) AND (1.26) AND PAINLEVÉ EQUATIONS OF II AND III TYPES

Let us start from equation (1.9). The deformation parameter here is only one - the variable x . Thus the isomonodromic deformation system (2.3) has to be reduced to a system of ordinary differential equations on two functions $u=u(x)$ and $w=w(x)$.

The form Ω now is

$$\Omega(\lambda) \equiv \Omega^{(\infty)}(\lambda) = U(\lambda)dx,$$

where $U(\lambda)$ is a polynom of first order in λ , defined according to (2.2) by equation

$$U(\lambda) = -i\lambda\sigma_3 + i[\sigma_3, \psi_1] .$$

Here ψ_1 is the first coefficient of formal series (1.11'). Taking into account formulae (1.11") and (1.12') we obtain an expression for $U(\lambda)$ through the coefficients of $A(\lambda)$ matrix:

$$U(\lambda) = -i\lambda\sigma_3 + \frac{1}{4}A_{-2} = -i\lambda\sigma_3 - u\sigma_2 .$$

The equations (2.3) and (2.4) are written now in the form

$$\frac{\partial\psi}{\partial x} = [-i\lambda\sigma_3 - u\sigma_2]\psi , \tag{3.1}$$

$$-i\sigma_3 - 4iu_x\sigma_3 - 4u_x\lambda\sigma_2 - 2w_x\sigma_1 = \tag{3.2}$$

$$= -i\sigma_3 + [-i\lambda\sigma_3 - u\sigma_2, -(4i\lambda^2 + ix + 2iu^2)\sigma_3 - 4u\lambda\sigma_2 - 2w\sigma_1].$$

Calculating the commutator in eq. (3.2) and scaling the terms of equal order in λ , it is easy to transform eq. (3.2) into the equations

$$w = u_x ,$$
$$u_{xx} = 2u^3 + xu. \tag{3.3}$$

The equation (3.3) is a particular case of the second Painlevé equation (where the constant ν in its right hand side is equal to zero). The result of our discussion is formulated in the following theorem.

THEOREM 3.1. (Flaschka-Newell). The smooth functions $u(x)$, $w(x)$ describe isomonodromic deformations of eq. (1.9) if $w=u_x$ and $u(x)$ satisfy Painlevé equation of the second type (3.3).

REMARK 3.1. In fact, more strong statement is proved in [16] . It was shown there that the general Painlevé II equation (with $\nu\neq 0$) describes isomonodromic deformations of eq. (1.9), where the regular singularity is added:

$$A(\lambda) \mapsto A(\lambda)+\frac{\nu}{\lambda}\sigma_2 .$$

Let us pass now to the system (1.26). The calculations similar to the performed above lead to expression for the Ω form:

$$\Omega(\lambda)=U(\lambda)dx ,$$

$$U(\lambda)=-\frac{ix\lambda}{8}\sigma_3+\frac{2}{x}A_{00}=$$

$$=-\frac{ix}{8}\lambda\sigma_3-\frac{i}{2}w\sigma_1 . \qquad (3.4)$$

Substituting $\Omega(\lambda)$ from eq. (3.4), and $A(\lambda)$ from (1.26) into isomonodromic deformation equation (2.4), we obtain equations:

$$w = u_x$$

$$u_{xx}+\frac{1}{x}u_x+\sin u = 0 . \qquad (3.5)$$

Making the change of variables

$$u \longrightarrow i\exp(iu/2), \quad x \rightarrow x\, 2^{-3/2}$$

we reduce the equation (3.5) to a particular case of Painlevé III equation ($\alpha = \beta = 0$).

Thus the following theorem is true.

THEOREM 3.2. (Flaschka-Newell). The smooth functions $u(x)$ and $w(u)$ describe isomodromic deformations of eq. (1.26) iff $w = u_x$ and $u(x)$ satisfy Painlevé III equation (3.5).

Chapter 4. INVERSE PROBLEM OF THE MONODROMY THEORY FOR THE SYSTEMS (1.9) AND (1.26). ASYMPTOTIC ANALYSIS OF INTEGRAL EQUATIONS OF THE INVERSE PROBLEM

The results obtained in previous chapters reduce the problem of integration of PII and $PIII$ equations to the solution of the monodromy theory inverse problem for the systems (1.9) and (1.26) respectively. The inverse problem in question consists of determination of values u, w through the given parameters p, q which are independent of x. Due to the theorems 3.1 and 3.2, u and w appear to be the smooth functions of x and produce the solutions of corresponding Painlevé equations.

Let us discuss therefore to what analitical problem the inverse problem of monodromy theory itself is reduced. We start with the system (1.9).

First we define the matrix-valued functions $\chi_k(\lambda)$, $k=1,\dots,6$ as follows

$$\chi_k(\lambda) = \Psi_k(\lambda)\cdot \exp\left\{\frac{4i}{3}\lambda^3\sigma_3 + ix\lambda\sigma_3\right\}, \quad \chi_7 \equiv \chi_1 , \tag{4.1}$$

where $\Psi_k(\lambda)$ are the canonical solutions of the system (1.9), defined in Chapter 1. The functions $\chi_k(\lambda)$ are in fact the entire functions of λ, tending to unity as $\lambda \to \infty$ in the corresponding sectors Ω_k. Moreover they are conjugated with one another by the equations

$$\chi_{k+1}(\lambda) = \chi_k(\lambda)\, G_k(\lambda, x),$$

where

$$G_k(\lambda,x) = \exp\left\{\frac{-4i}{3}\lambda^3\sigma_3 - ix\lambda\sigma_3\right\} S_k \cdot$$

$$\cdot \exp\left\{+\frac{4i}{3}\lambda^3\sigma_3 + ix\lambda\sigma_3\right\} . \tag{4.2}$$

The properties of χ_k mentioned above may be stated as follows

1. $\chi_k(\lambda)$ are analitical in ω_k and continuous in $\overline{\omega}_k$, where

$$\omega_k = \left\{\lambda : \frac{\pi}{3}(k-1) < \arg\lambda < \frac{\pi}{3}k\right\}, \quad k = 1, 2, \ldots, 6 .$$

2. the conjugation equations hold on the rays $\Gamma_k = \left\{\lambda : \arg\lambda = \frac{\pi k}{3}\right\}$:

$$\chi_{k+1}(\lambda) = \chi_k(\lambda) G_k(\lambda), \quad \lambda \in \Gamma_k .$$

3. the asymptotics take place

$$\chi_k(\lambda) \to I, \quad \lambda \to \infty, \quad \lambda \in \tilde{\omega}_k$$

in the semi-open sectors

$$\tilde{\omega}_k = \left\{\lambda : \frac{\pi}{3}(k-1) \leq \arg\lambda < \frac{\pi}{3}k\right\} .$$

In other words the functions $\chi_k(\lambda)$ being constructed through the solution of system (1.9) present the solution of regular ($\det \chi_k(\lambda) \neq 0$ for any λ, k) matrix-valued Riemann-Hilbert (RH) problem on the family of rays Γ_k with the conjugation matrices $G_k(\lambda)$ of special form (4.2). Moreover, the inverse statement is true. Namely, we introduce the matrices $G_k(\lambda)$ by formulae (4.2), (1.22) and consider the RH problem 1 - 3. Suppose that it has a solution, then the functions

$$\Psi_k(\lambda) = \chi_k(\lambda) \exp\left\{-\frac{4i}{3}\lambda^3\sigma_3 - ix\lambda\sigma_3\right\}$$

would have the asymptotics (1.14) in the sectors $\tilde{\omega}_k$. The conjugation conditions hold on the boundaries of the sectors

$$\Psi_{K+1} = \Psi_K S_K ,$$

with matrices S_K being independent of λ. The latter circumstance imply evident equalities

$$(\Psi_1)_\lambda \Psi_1^{-1} = \ldots = (\Psi_K)_\lambda \Psi_K^{-1} = \ldots = (\Psi_6)_\lambda \Psi_6^{-1} .$$

Applying now the Lioville's theorem together with the asymptotics (1.14), we obtain immediately that all Ψ_K are just the solutions of the same differential equation

$$\frac{d\Psi}{d\lambda} = \left[-4i\lambda^2 \sigma_3 + A_1\lambda + A_0\right]\Psi . \tag{4.3}$$

The evident reduction constraints (1.18) on S_K matrices are at once transformed into corresponding constraints (1.17) on Ψ_K functions, which in its turn yields that the structure of coefficients A_0, A_1 in eq. (4.3) coinside with those of eq. (1.9). Summing up we obtain the following theorem.

THEOREM 4.1. (Flaschka-Newell [16]). The inverse problem of the monodromy theory for the system (1.9) is equivalent to the matrix RH problem 1 - 3, where the conjugation matrices are defined by the formulae (4.2), (1.22). The coefficients of the system (1.9) are reconstructed through the solution of RH problem as follows

$$u = 2(\psi_1)_{12} ,$$

$$w = -2i(\psi_2)_{12} - 2i([\psi_1, \sigma_3]\psi_1)_{12} , \tag{4.4}$$

$$\psi_1 = \lim_{\substack{\lambda\to\infty \\ \lambda\in\omega_K}} \lambda \chi_K(\lambda) , \quad \psi_2 = \lim_{\substack{\lambda\to\infty \\ \lambda\in\omega_K}} \lambda^2 \chi_K(\lambda).$$

The interesting paper by M.Ablowitz and A.Focas [37] shows that the solution of the RH problem 1 - 3 on the six rays Γ_k might be reduced to the sequence of three RH problems on the straight line *) To our regret this transformation is quite insufficient in order to obtain the simple criteria of solvability of the initial problem. Hence the complete investigation of the solvability of the RH problem 1 - 3 appears to be nowaday an open problem. We are also unable to present an exhaustive description of the behaviour of the solution $u(x)$ to $P\text{II}$ equation in terms of the monodromy parameters p and q. Any constructive results are achievable so far only in a particular case, to the description of which we are now passing.

Assume that the monodromy data for the system (1.9) satisfy the equation

$$p = -q \quad , \tag{4.5}$$

which implies the triviality of the Stokes matrices S_2 and S_5 :

$$S_2 = S_5 = I \quad . \tag{4.6}$$

That means in its turn the applicability of the canonical asymptotics (1.14) for Ψ_2 and Ψ_5 in the open upper and lower half planes respectively. The conjugation equation takes place

$$\begin{gathered} \Psi_5 = \Psi_2 G_0 \quad , \\ G_0 = S_3 S_4 = S_1^{-1} S_6^{-1} = \begin{pmatrix} 1 & -p \\ -p & 1+p^2 \end{pmatrix} . \end{gathered} \tag{4.7}$$

---

*) The general case of $P\text{II}$ equation with $\nu \neq 0$ is also considered in ref. [37] . The corresponding RH problem in this case appears to be discontinuous and singular at the origin, i.e. the branching of $\Psi_k(\lambda)$ occurs as $\lambda \to 0$.

Introducing the piecewise analytic function

$$\chi(\lambda)=\begin{cases}\Psi_2(\lambda)\exp\left\{\frac{4i}{3}\lambda^3\sigma_3+ix\lambda\sigma_3\right\}, & \operatorname{Im}\lambda\geqslant 0,\\ \Psi_5(\lambda)\exp\left\{\frac{4i}{3}\lambda^3\sigma_3+ix\lambda\sigma_3\right\}, & \operatorname{Im}\lambda\leqslant 0,\end{cases}$$

we conclude that it solves the following Riemann-Hilbert problem on the real line:

1. $\chi(\lambda)$ is analytical as $\operatorname{Im}\lambda>0$ or $\operatorname{Im}\lambda<0$.
2. $\chi^-(\lambda)=\chi^+(\lambda)G(\lambda)$, $\operatorname{Im}\lambda=0$,

$$G(\lambda)=\exp\left\{-\frac{4i}{3}\lambda^3\sigma_3-ix\lambda\sigma_3\right\}G_0\exp\left\{\frac{4i}{3}\lambda^3\sigma_3+ix\lambda\sigma_3\right\}=$$

$$=\begin{pmatrix}1 & -pee^{-\frac{8}{3}i\lambda^3-2ix\lambda}\\ -pe^{\frac{8i}{3}\lambda^3+2ix\lambda} & 1+p^2\end{pmatrix}.$$

3. $\chi(\lambda)\to I$ as $\lambda\to\infty$, $\operatorname{Im}\lambda\neq 0$.

Applying the standard reasoning (see, for example, [37]) we may write out the singular integral equation, equivalent to the problem 1 - 3. It has the form

$$\chi^+(\lambda)=I+\frac{1}{2\pi i}\int_{-\infty}^{\infty}\frac{1}{\eta-\lambda-i0}\chi^+(\eta)\left(I-G(\eta)\right)d\eta\,. \tag{4.8}$$

The equation (4.8) plays the same role in monodromy theory inverse problem for the system (1.9) (with limitation (4.5) on monodromy data) as the Gelfand-Levitan-Marchenko equation in the inverse scattering problem. However in comparison with the latter it is studied much less extensively. Strictly speaking nowaday there is only one precise result has been published in concern of this problem

THEOREM 4.2. ([34]). Suppose alongside with (4.5) that the monodromy parameter ρ being purely imaginary:

$$\rho = ia\ , \quad a \in \mathbb{R}\ ,$$

which guarantees the solution of P_{II} equation $u(x)$ being real-valued. Then the two statements are true

1. If $|a| < 1$, then equation (4.8) is uniquely solvable for any real-valued x . The corresponding solution $u(x)$ of P_{II} equation appears to be smooth function for all $x \in \mathbb{R}$ and it is exponentially decreasing as $x \to +\infty$:

$$u(x) = \frac{i\rho}{2\sqrt{\pi}}\, x^{-1/4} e^{-2/3 x^{3/2}} \bigl(1 + o(1)\bigr) \ . \tag{4.9}$$

2. If $|a| > 1$, then there exists at least one real-valued $x = x_0$ for which the equation (4.8) is not solvable uniquely. The corresponding $u(x)$ has a pole in the point $x = x_0$.

From the viewpoint of our aims in this work the most interesting statement of theorem 4.2 is about the asymptotic behaviour of $u(x)$ as $x \to +\infty$. Let us show how the formula (4.9) might be derived from the integral equation (4.8) under assumption of its solvability. For this purpose we first rescale the variables in equation (4.8):

$$\begin{aligned} \lambda &\mapsto z = \lambda \cdot x^{-1/2} \\ \eta &\mapsto \xi = \eta \cdot x^{-1/2} . \end{aligned} \tag{4.10}$$

Retaining the old denotation for the function in (4.8), it is easy to write out the system of scalar integral equations on the matrix elements

$\chi_{21}^{+}(z)$, $\chi_{22}^{+}(z)$:

$$\chi_{21}^{+}(z)=\frac{\rho}{2\pi i}\int_{-\infty}^{\infty}\frac{\chi_{22}^{+}(\xi)e^{ix^{3/2}\theta(\xi)}}{\xi-z-i0}\,d\xi\ ,$$

$$\chi_{22}^{+}(z)=1+\frac{\rho}{2\pi i}\int_{-\infty}^{\infty}\frac{\chi_{21}^{+}(\xi)e^{-ix^{3/2}\theta(\xi)}}{\xi-z-i0}\,d\xi-\frac{\rho^{2}}{2\pi i}\int_{-\infty}^{\infty}\frac{\chi_{22}^{+}(\xi)}{\xi-z-i0}\,d\xi\ , \tag{4.11}$$

where $\theta(\xi)=\frac{8}{3}\xi^{3}+2\xi$.

The solution $u(x)$ of $P\,\text{II}$ equation may be expressed due to (4.4) and the change of variables (4.10) in the form

$$u(x)=-x^{1/2}\frac{\rho}{\pi i}\int_{-\infty}^{\infty}\chi_{22}^{+}(\xi)e^{ix^{3/2}\theta(\xi)}\,d\xi\ . \tag{4.12}$$

We need the asymptotics of $\chi_{22}^{+}(\xi)$ as $x\to+\infty$, which can be derived directly from the system (4.11). The latter is just the system of linear singular integral equations with rapidly oscillating kernels. Hence the decisive role for asymptotic investigation of the solutions plays the behaviour of the derivative $\theta'(\xi)$ on the real axis. We have

$$\theta'(\xi)=8\xi^{2}+2>0\ ,\quad -\infty<\xi<\infty\ . \tag{4.13}$$

This inequality means (see, for example, [52]) that for any smooth function $f(\xi)$ the estimate holds uniformly in $z\in\mathbb{R}$:

$$\frac{1}{2\pi i}\int_{-\infty}^{\infty}\frac{f(\xi)}{\xi-z-i0}\,e^{ix^{3/2}\theta(\xi)}\,d\xi=$$

$$= f(z)e^{ix^{3/2}\theta(z)} + o(1)\,, \quad x \longrightarrow +\infty\,. \tag{4.14}$$

In the case when $f(z)$ is analytically expandable into upper half-plane and tends to a constant at infinity the estimate (4.14) may be strenthened. Namely, deforming the contour of integration, so that it passes through the stationary-phase point $i/2$, we obtain

$$\frac{1}{2\pi i}\int_{-\infty}^{\infty} \frac{f(\xi)}{\xi - z - i0}\, e^{ix^{3/2}\theta(\xi)}\, d\xi =$$

$$= f(z)e^{ix^{3/2}\theta(z)} + \frac{1}{4i\sqrt{\pi}}\,\frac{f(i/2)}{i/2 - z}\, x^{-3/4} e^{-2/3 x^{3/2}}\left(1 + o(1)\right), \tag{4.15}$$

$$x \longrightarrow \infty\,, \quad z \in \mathbb{R}\,.$$

Suppose now that function $\chi^{+}_{22}(z)$ contains no oscillations in the leading order in x. Then the first equation (4.11) together with formula (4.15) yield

$$\chi^{+}_{21}(z) = \rho\,\chi^{+}_{22}(z)e^{ix^{3/2}\theta(z)} +$$

$$+ \frac{\rho}{4i\sqrt{\pi}}\,\frac{\chi^{+}_{22}(z)}{i/2 - z}\, x^{-3/4} e^{-2/3 x^{3/2}}\left(1 + o(1)\right), \quad x \longrightarrow +\infty\,,$$

uniformly in $z \in \mathbb{R}$. Substituting this equality into the second equation of the system (4.11) we conclude that integrals cancel in the leading order of right-hand side, and an asymptotic representation for $\chi^{+}_{22}(z)$ occurs

$$\chi^{+}_{22}(z) = 1 - \frac{\rho^2}{8\pi\sqrt{\pi}}\, x^{-3/4} e^{-2/3 x^{3/2}}\, \chi^{+}_{22}(i/2)\left(I(z,x) + o(1)\right)\,,$$

$$x \longrightarrow +\infty ,$$

where

$$I(z,x) = \int_{-\infty}^{\infty} \frac{e^{-ix^{3/2}\theta(\xi)}}{(i/2-\xi)(\xi-z-i0)}\, d\xi .$$

The asymptotic evaluation of this integral demands, according to the stationary-phase method, to deformate the contour of integration into lower half-plane, where there is no singularities of integrand:

$$I(z,x) = O(x^{-3/4} e^{-2/3 x^{3/2}} \frac{1}{z+i/2}) .$$

We arrive thus to the exponentially decreasing asymptotics for $\chi^{+}_{22}(z)$ which is uniform in $z \in \mathbb{R}$:

$$\chi^{+}_{22}(z) = 1 + o(x^{-3/4} e^{-2/3 x^{3/2}}) . \tag{4.16}$$

This asymptotics correlated with an apriori hypothesis about the absense of oscillations in $\chi^{+}_{22}(z)$, and, making use of (4.12), it yields the asymptotic representation for $u(x)$

$$u(x) =$$

$$= -x^{+1/2} \frac{\rho}{\pi i} \int_{-\infty}^{\infty} e^{ix^{3/2}\theta(\xi)} d\xi + o(x^{-1/4} e^{-2/3 x^{3/2}}) . \tag{4.17}$$

The evaluation of integral in rhs of (4.17) with the help of stationary-phase method lead directly to the formula (4.9).

REMARK 4.1. It is evident that the leading term of asymptotics (4.17) is at the same time the leading term in ρ , assuming ρ to be small. Moreover the expression (4.17) is precisely the integral representation of Airy function $A_i(x)$, arising as a solution of

Airy equation solved by the Laplace method. Note that Airy equation is just linearisation of P_{II} equation. Therefore we may conclude that in the limit case of small monodromy data the method of isomonodromic deformations reduces to the Laplace's method of solving a corresponding linearised equation. It is quite similar to the relationship between the method of inverse scattering problem for partial nonlinear equations and Fourier method. The latter may be treated as a limit case of the inverse scattering method when the scattering matrix is nearly trivial.

Looking back at asymptotic analysis of the system (4.11) it is easy to learn the essence of difficulties which would occur in any attempt to construct the asymptotics of solution as $x \to -\infty$ in the same manner as above. In fact, replacing x on $-x$ and $\theta(\xi) = \frac{8}{3}\xi^3 + 2\xi$ on $\tilde{\theta}(\xi) = \frac{8}{3}\xi^3 - 2\xi$ in rhs of (4.11), (4.12), we obtain the case of large negative x. The essential difference now is attributed to the new position of stationary-phase points $\xi = \pm\frac{1}{2}$:

$$\tilde{\theta}'(\xi) = 8\xi^2 - 2 \Rightarrow \tilde{\theta}'(\pm\tfrac{1}{2}) = 0 .$$

As a result the structure of estimate (4.14) is changed at once — it fails now to be uniform in $z \in \mathbb{R}$:

$$\frac{1}{2\pi i}\int_{-\infty}^{\infty} \frac{f(\xi)}{\xi - z - i0} e^{i(-x)^{3/2}\tilde{\theta}(\xi)} d\xi =$$

$$= f(z) e^{i(-x)^{3/2}\tilde{\theta}(z)} \eta(z^2 - \tfrac{1}{4}) + O\left(\frac{1}{(-x)^{3/4}(z^2 - 1/4)}\right) , \qquad (4.18)$$

$$\eta(z) = \begin{cases} 1 , & z > 0 , \\ 0 , & z < 0 . \end{cases}$$

Let us ignore for a while this non-uniformness. Then, according to our scheme developed above in the case of $x \to +\infty$, we get the representation for $\chi_{21}^{+}(z)$:

$$\chi_{21}^{+}(z) \approx \rho \chi_{22}^{+}(z) e^{i(-x)^{3/2}\theta(z)} \eta(z^2 - \tfrac{1}{4}), \quad x \to -\infty .$$

The substitution of this expression into the second equation (4.11) leads now not to cancellation of integrals in right-hand side of it, but to the singular integral equation

$$\chi_{22}^{+}(z) = 1 - \frac{\rho^2}{2\pi i} \int_{-1/2}^{1/2} \frac{\chi_{22}^{+}(\xi)}{\xi - z - i0} d\xi ,$$

which might be treated as an integral equation equivalent to the following scalar Riemann-Hilbert problem

$$\chi_{22}^{-}(z) = \chi_{22}^{+}(z)\left(1 + \rho^2 \eta(1/4 - z^2)\right), \quad \operatorname{Im} z = 0,$$
$$\chi_{22}(\infty) = 1. \tag{4.19}$$

The solution of the problem (4.19) is written out easily:

$$\chi_{22}(z) = \left(\frac{z - \frac{1}{2}}{z + \frac{1}{2}}\right)^{i\nu}, \quad \nu = \frac{1}{2\pi} \ln(1 + \rho^2) , \tag{4.20}$$

where we take the principal branch of corresponding multi-valued function with the interval $\left(-\frac{1}{2}, \frac{1}{2}\right)$ as a cut.

We'll assert again $\rho = ia$, $|a| < 1$, which guarantee the existence of smooth real-valued solution $u(x)$ of $P\,\mathrm{II}$ equation. Under this assumption we have $\nu \in \mathbb{R}$ and thus substituting (4.20) into (4.12) and evaluating integrals by the stationary-phase method, we arrive to the following asymptotics for $u(x)$:

$$u(x) \approx \alpha \cdot (-x)^{-1/4} \cos\left\{ \frac{2}{3}(-x)^{3/2} + \beta \ln(-x) + \varphi \right\} , \tag{4.21}$$

where

$$\begin{aligned} \alpha &= -\frac{a}{\pi} \cdot \frac{1}{2} e^{-\pi\nu/4} (1 + e^{-\pi\nu}) \left| \Gamma\left(\frac{i\nu}{2} + \frac{1}{2}\right) \right| , \\ \beta &= \frac{3}{4}\nu , \\ \varphi &= -\frac{\pi}{4} + \nu \ln 2 - \arg \Gamma\left(\frac{i\nu}{2} + \frac{1}{2}\right) , \\ \nu &= \frac{1}{2\pi} \ln(1 - a^2) . \end{aligned} \tag{4.22}$$

In spite of qualitative valueness of this asymptotics, we easily check that it is not true. Namely, the direct substitution of (4.21) into P_{II} equations yields with necessity that

$$\alpha^2 = -\frac{4}{3}\beta , \quad \beta < 0 .$$

At the same time the constraint (4.22) provide another equation

$$\alpha^2 = -\frac{1}{\pi} \operatorname{sh} \frac{4\pi}{3} \beta \cdot \left(1 + e^{-\pi \frac{4}{3}\beta}\right) , \quad \beta < 0 ,$$

which contradicts the previous one.

The reason of our failure with asymptotics (4.21), (4.22) lies in our permission to ignore the singularity of estimate (4.18) near the stationary points $\xi = \pm \frac{1}{2}$. It is clear now that we have to establish the precise asymptotics of $\chi^{+}_{22}(\xi)$ near the stationary points $\xi = \pm \frac{1}{2}$ in order to derive the asymptotics of $u(x)$ as $x \longrightarrow -\infty$ from the integral

$$u(x) = -(-x)^{1/2} \frac{p}{\pi i} \int_{-\infty}^{\infty} \chi^{+}_{22}(\xi) e^{i(-x)^{3/2}\tilde{\theta}(\xi)} d\xi .$$

However the demonstrated above method of treatment of integral equations (4.11) (or Riemann-Hilbert problem 1 - 3) does not provide in principle the correct asymptotics of $\chi^{+}_{22}(z)$ as $x \to -\infty$ and $z \sim \pm 1/2$.

Let us discuss now the inverse problem of the monodromy theory for the system (1.26), which is equivalent, according to theorem 3.1, to integration of $P\,III$ equation. We are able here to formulate the RH problem on the real axis for the general case without any constraints on the monodromy data p and q.

We define the pair of functions $\Phi_1(\lambda)$ and $\Phi_2(\lambda)$ in open upper and lower half-planes respectively:

$$\Phi_1(\lambda) = \Psi_+(\lambda), \quad \Phi_2(\lambda) = \Psi_-(\lambda), \tag{4.23}$$

where the functions $\Psi_+(\lambda)$ and $\Psi_-(\lambda)$ are defined (see the Chapter 1) respectively in the domains $0 < \arg\lambda < \pi$ and $-\pi < \arg\lambda < 0$. The continuation of $\Phi_1(\lambda)$ and $\Phi_2(\lambda)$ onto the real axis is produced by the formulae (1.43), (1.49) as follows

$$\begin{aligned} \lambda > 0: \quad & \Phi_1(\lambda) = \Psi_1^{(\infty)}(\lambda) \cdot S_+^{(\infty)}, \\ & \Phi_2(\lambda) = \Psi_1^{(\infty)}(\lambda) \cdot S_-^{(\infty)}, \end{aligned} \tag{4.24}$$

$$\begin{aligned} \lambda < 0: \quad & \Phi_1(\lambda) = \Psi_2^{(\infty)}(\lambda) \cdot \sigma_1 \cdot S_-^{(\infty)} \sigma_1, \\ & \Phi_2(\lambda) = \Psi_2^{(\infty)}(\lambda) \cdot \sigma_1 \cdot S_+^{(\infty)} \sigma_1. \end{aligned} \tag{4.25}$$

As a result an additional singularity (besides the essential singularity at the origin) of $\Phi_{1,2}(\lambda)$ occurs at $\lambda=0$. Nevertheless, we have the cnjugation equation on the real axis

$$\Phi_2(\lambda)=\Phi_1(\lambda)G_0(\lambda),\quad Im\lambda=0,$$

$$G_0(\lambda)=\begin{cases}\begin{pmatrix}1 & -p\\ -q & 1+pq\end{pmatrix}, & \lambda>0,\\ \begin{pmatrix}1 & q\\ p & 1+pq\end{pmatrix}, & \lambda<0.\end{cases} \qquad (4.26)$$

Taking into account the asymptotic behaviour of $\Phi_{1,2}(\lambda)$ at the singular points $\lambda=0$, $\lambda=\infty$ (see the formula (1.41)), we introduce, just in the same manner as above, the piecewise analytical function

$$\chi(\lambda)=\begin{cases}\Phi_1(\lambda)\exp\left\{\frac{ix^2}{16}\lambda\sigma_3+\frac{i}{\lambda}\sigma_3\right\}, & Im\lambda\geq 0,\\ \Phi_2(\lambda)\exp\left\{\frac{ix^2}{16}\lambda\sigma_3+\frac{i}{\lambda}\sigma_3\right\}, & Im\lambda\leq 0.\end{cases}$$

It permits us to reformulate the inverse problem for the system (1.26) in the form of RH problem on the real axis:

1'. $\chi(\lambda)$ is analytical function as $Im\lambda>0$ and $Im\lambda<0$.

2'. $\chi^-(\lambda)=\chi^+(\lambda)G(\lambda,x)$, $Im\lambda=0$,

where

$$G(\lambda,x)=$$

$$=\exp\left\{-\frac{ix^2}{16}\lambda\sigma_3-\frac{i}{\lambda}\sigma_3\right\}G_0(\lambda)\exp\left\{\frac{ix^2}{16}\lambda\sigma_3+\frac{i}{\lambda}\sigma_3\right\},$$

and $\sigma_0(\lambda)$ is defined by (4.26).

3'. $\chi(\lambda) \longrightarrow I$ as $\lambda \longrightarrow \infty, \ Im\lambda \neq 0$.

In a general case the problem 1' - 3' is discontinuous in such a sense that $\sigma_0(\lambda)$ matrix is piecewise constant to the right and to the left of the origin. However, under the assumption $p = -q$ (the same as for the system (1.9)!) the problem 1' - 3' in question appears to be regular, and we are able to reduce it to singular integral equation quite similar to that of (4.8):

$$\chi^+(\lambda) = I + \frac{1}{2\pi i} \int_{-\infty}^{\infty} \frac{1}{\eta - \lambda - i0} \chi^+(\eta) \times$$
$$\times \begin{pmatrix} 0 & p e^{-\frac{ix^2}{8}\eta - \frac{2i}{\eta}} \\ -p e^{\frac{ix^2}{8}\eta + \frac{2i}{\eta}} & p^2 \end{pmatrix} d\eta \ . \tag{4.27}$$

Let us try to analyse this equation in order to extract an asymptotics of $u(x)$ - the corresponding solution of $P\,\text{III}$ equation as $x \longrightarrow \pm\infty$. We'll proceed in the same way as above for the case of equation (4.8). The variables are rescaled here independently of the sign of x :

$$\lambda \longmapsto z = \lambda \cdot x \ ,$$

$$\eta \longmapsto \xi = \eta \cdot x$$

This implies the appearence of rapid oscillations in the kernel of equation (4.27), where the phase takes the form

$$x \cdot \left(\frac{\xi}{8} + \frac{2}{\xi} \right) .$$

Hence the stationary-phase points are $\xi = \pm 4$ and so, independently of the sign of x , we find ourselves in a situation when the stationary point always lie on the contour of integration. Thus the singularities arise in the asymptotics of $\chi^{+}(\lambda)$ near the stationary points, which makes impossible to construct in a simple manner the required asymptotics of $u(x)$.

The analysis of the inverse problems of monodromy theory for the systems (1.9) and (1.26) produced in this chapter, reveals rather limited capabilities of their application to the asymptotic studies of solutions to $P\,\underline{\overline{II}}$ and $P\,\underline{\overline{III}}$ equations. A much more effective approach seems to deal with direct problem of the monodromy theory. The concise exposition of this method would be presented in the forthcoming chapters 5 - 9.

Chapter 5. ASYMPTOTIC SOLUTION TO A DIRECT PROBLEM OF THE MONODROMY THEORY FOR THE SYSTEM (1.9)

The main purpose of this Chapter is to prove the following theorem.

THEOREM 5.1. Let the parameters u, w of the system (1.9) depend on x in such a way that

a)
$$u = O((-x)^{-1/4}),$$
$$w = O((-x)^{+1/4}), \quad x \to -\infty, \tag{5.1}$$

b) at least one of the values v, v^*,

$$v = \sqrt{2i}(-x)^{3/4}\left[\frac{i}{2}(-x)^{-1/2}u - \frac{w}{2}(-x)^{-1}\right],$$

$$v^* = -\sqrt{2i}(-x)^{3/4}\left[\frac{i}{2}(-x)^{-1/2}u + \frac{w}{2}(-x)^{-1}\right]$$

is separated from zero as $x \longrightarrow -\infty$,

c) $-1 \leqslant \operatorname{Re} vv^* \leqslant 0$.

Then the monodromy data p, q have the following asymptotic expansions

$$p = \frac{i}{v}\frac{\sqrt{2\pi}}{\Gamma(i\delta)}\exp\left\{\frac{2}{3}i(-x)^{3/2} + i\delta\ln(-x)^{3/2} + i\delta\ln 8 + \frac{\pi}{2}\delta\right\}(1+o(1)),$$

$$q = \frac{1}{v^*}\cdot\frac{\sqrt{2\pi}}{\Gamma(-i\delta)}\exp\left\{-\frac{2}{3}i(-x)^{3/2} - i\delta\ln(-x)^{3/2} - i\delta\ln 8 + \right.$$

$$\left. + \frac{\pi}{2}\delta\right\}(1+o(1)),$$

$$i\delta = -vv^*, \quad x \longrightarrow -\infty. \tag{5.2}$$

Before getting down to the proof of the theorem 5.1 let us outline its crusical points.

For our asymptotic study of the system (1.9) as $x \to -\infty$ it is convenient to scale the variable λ :

$$z = (-x)^{-1/2} \lambda$$

(compare with (4.10)). After this change of variables the system (1.9) takes a new form

$$\frac{d\Psi}{dz} = \tau A_0(z,\tau)\Psi , \tag{5.3}$$

where

$$\tau = (-x)^{3/2} ,$$

$$A_0 = -(4iz^2 - i - \frac{2iu^2}{x})\sigma_3 - 4uz(-x)^{-1/2}\sigma_2 + \frac{2w}{x}\sigma_1 .$$

In eq. (5.3) τ is large parameter and for the matrix $A_0(z,\tau)$ according to (5.1), an estimate holds

$$A_0(z,\tau) = O(1) , \quad \tau \to +\infty ,$$

where z is fixed. In other words we get here the situation typical to WKB-method. According to its ideology it is neccesary to figure out the turning points of eq. (5.3) and then to seek an asymptotic solution for away from these points in the form

$$\Psi_{WKB} \simeq T(z) exp\{\tau \int^{z} \Lambda(\eta) d\eta\} \tag{5.4}$$

where $\Lambda(z)$ is diagonalization of A_0 and T conjugates the matrices A_0, Λ :

$$\Lambda = T^{-1} A_0 T .$$

As the matrix $A_0(z)$ is polynomial of z the WKB-approximation

(5.4) turns out to be double (see [38]), i.e. it is simultaneously an asymptotics as $z \to \infty$. Thus we are able to compute the main term in τ of the matrices $C_K(\tau)$ linking the canonical solutions $\Psi_K(z)$ with WKB-solutions $\Psi^K_{WKB}(z)$ in their corresponding sectors Ω_K :

$$\Psi_K = \Psi_{WKB} \cdot C_K \; .$$

Furthermore due to the double nature of WKB-approximation the formula (5.4) for the solutions Ψ^K_{WKB} goes to be true in some circular domain near the turning points. In the neighbourhood of the turning points our initial system may be simplified and is reduced to exactly solvable system. As would be shown below the system (5.3) has two multiple turning points $z_\pm = \pm \frac{1}{2}$ *) in the neighbourhood of which the solution $\overset{\circ}{\Psi}_\pm$ of the system (5.3) may be expressed through the parabolic cylinder functions. In the circular domain mentioned above the solution $\overset{\circ}{\Psi}_+(z)$ may be matched with WKB-solutions Ψ^K_{WKB} , $K=1,2,\ K=6$. That means the possibility of calculation of matrices N_K such that

$$N_K = \overset{\circ}{\Psi}{}^{-1}_+ \Psi^K_{WKB} \qquad (5.5)$$

in the leading order of τ .

The monodromy data for the system (1.9) i.e. the parameters p, q may be constructed by the Stokes matrices S_1 and S_6. From eq. (5.4) and eq. (5.5) we have

---

*) It is worth mentioning that these turning points coinside with stationary phase points in the kernels of integral equations (4.8).

$$S_1 = C_1^{-1} N_1^{-1} N_2 C_2 \quad , \tag{5.6}$$

$$S_6 = C_6^{-1} N_6^{-1} N_1 C_1 \quad . \tag{5.7}$$

Thus we can express directly from here the leading terms in τ of the parameters p, q .*)

Summing up we present the scheme of the theorem's 5.1 proof comprising the following points:

I) Construction of WKB-solutions Ψ^{K}_{WKB} , $K = 1, 2, 6$.

II) Construction of the solution $\mathring{\Psi}_{+}(z)$ at the neighbourhood of the turning point.

III) Computation of the leading term in τ for the matrices C_K.

IV) Computation of the leading term in τ for the matrices N_K.

V) Computation of p and q by eq. (5.6), (5.7) as a final step of the proof.

Let us pass to the concrete realization of this programme.

I. WKB-SOLUTIONS.

The eigenvalues $\mu_{1,2}(z, \tau)$ of the matrix $A_0(z,\tau)$ are given by equations

$$\mu_{1,2} = \pm \mu \ ,$$

$$\mu = 4i\sqrt{(z^2 - \frac{1}{4})^2 + \frac{u^2}{4x} + \frac{u^4}{4x^2} - \frac{w^2}{4x^2}} \quad . \tag{5.8}$$

---

*) Certaining it is possible to obtain full asymptotic expansions for p and q by computting appropriate number of terms for matrices C_K and N_K . However, for our purpose - the construction of asymptotic solutions of Painlevé equations - it is sufficient to know only the leading order terms of p, q .

The condition $\mu = 0$ defines turning points. Apriori there are four of them:

$$z_{1,2} = \sqrt{\frac{1}{4} \pm \sqrt{\frac{w^2}{4x^2} - \frac{u^2}{4x} - \frac{u^4}{4x^2}}} \ ,$$

$$z_{3,4} = -\sqrt{\frac{1}{4} \pm \sqrt{\frac{w^2}{4x^2} - \frac{u^2}{4x} - \frac{u^4}{4x^2}}} \ .$$

However, due to eq. (5.1) z_1 and z_2 tend to one point as $\tau \to \infty$ the same is true for z_3 and z_4 :

$$z_{1,2} \longrightarrow z_+ = \frac{1}{2} \ ,$$

$$z_{3,4} \longrightarrow z_- = -\frac{1}{2} \ , \quad \tau \to \infty \ .$$

The speed of convergence here is of order $\tau^{1/2}$, thus it is natural to apply WKB-asymptotics (5.4) in the domains where

$$\tau^{1/2} \left| (z \pm \tfrac{1}{2}) \right| \longrightarrow \infty \ . \tag{5.9}$$

It is easy to prove that matrices $\Lambda(z,\tau), T(z,\tau)$ for the system (5.3) may be defined as

$$\Lambda = -\mu \sigma_3 \ ,$$

$$T = \frac{1}{8i(z^2 - \frac{1}{4})} \left[(4iz^2 - i - 2i\frac{u^2}{x} + \mu)\sigma_0 + \frac{4izu}{(-x)^{1/2}} \sigma_1 + \right.$$

$$\left. + \frac{2iw}{x} \sigma_2 \right] , \quad \sigma_0 = I \ . \tag{5.10}$$

The function $\mu(z,\tau)$ has an asymptotics, derived from eq.(5.8) and condition a) in Theorem 5.1

$$\mu = 4i(z^2 - \tfrac{1}{4}) + O\left(\frac{1}{\tau(z^2 - \frac{1}{4})} \right) .$$

Thus, in a domain where eq. (5.9) holds, the following statements are true

1) $$T(z,\tau) = I + O\left(\frac{z}{\sqrt{\tau}(z^2-1/4)}\right) + O\left(\frac{1}{\sqrt{\tau}(z^2-1/4)}\right),$$

2) $$\int_z^{\infty} R(\eta,\tau)\,d\eta = O\left(\frac{1}{\sqrt{\tau}(z^2-1/4)}\right) + O\left(\frac{1}{\sqrt{\tau}} \ln \frac{z-1/2}{z+1/2}\right),$$

where $R = T^{-1}\dfrac{dT}{dz}$ is a remainder, occuring after substitution of (5.4) into eq. (5.3).

3) $\int^{z} \mu(\eta,\tau)\,d\eta \simeq \frac{4}{3} i z^3 - i z$ implies that conjugate Stokes lines $\left(\mathrm{Re} \int_{z_{1,2,3,4}}^{z} \mu(\eta,\tau)\,d\eta = 0\right)$ asymptotically tend to the real axis and to the hyperbola lines

$$\eta^2 = 3\left(\xi^2 - \frac{1}{4}\right), \quad z = \xi + i\eta$$

(see figure 5.1)

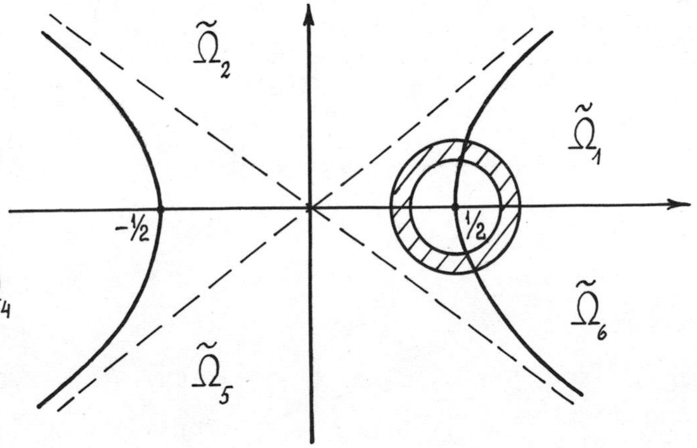

Figure 5.1

Figure 5.1. The solid lines designate the limit configuration of conjugate Stokes lines, the dotted lines - the rays $\arg z = \frac{\pi K}{3}$, shaded circular domain - the matching domain Ω^{ε}_{+} .

Applying the standard WKB-technique of integral equations (see, for example, [52]), it is easy to prove the following lemma

LEMMA 5.1. Let $\tilde{\Omega}_K$, $K = 1, 2, \ldots, 6$, be nonintersecting open domains, bounded by intervals $\left[-\frac{1}{2}, \frac{1}{2}\right], \left(-\infty, -\frac{1}{2}\right], \left[\frac{1}{2}, +\infty\right.$ and hyperbola lines $\eta^2 = 3(\xi^2 - \frac{1}{4})$, such that $\bigcup_K \overline{\tilde{\Omega}}_K = \mathbb{C}$ (see figure 5.1).

Then we have

a) in any domain $\tilde{\Omega}_K$ the equation (5.3) has a solution Ψ^K_{WKB} with double asymptotics of the form

$$\Psi^K_{WKB}(z,\tau) = \left[I + O\left(\frac{z}{\sqrt{\tau}(z^2 - 1/4)}\right) + O\left(\frac{1}{\sqrt{\tau}(z^2 - 1/4)}\right) + \right.$$

$$\left. + O\left(\frac{1}{\sqrt{\tau}} \ln \frac{z - 1/2}{z + 1/2}\right)\right] \exp\left\{-\tau\sigma_3 \int_{1/2}^{z} \mu(\eta,\tau)\,d\eta\right\},$$

$$|\tau^{1/2}(z^2 - 1/4)| \to \infty . \tag{5.11}$$

b) the asymptotics (5.11) retains as the solution Ψ^K_{WKB} is expanded to the domain $\tilde{\Omega}_{K+1}$, or to $\tilde{\Omega}_{K-1}$.

Later on we would be interested only in the solutions Ψ^K_{WKB} with $K = 1, 2$ and $K = 6$. Let us assume that they are expanded to the domains $\tilde{\Omega}_{K-1}$ ($\tilde{\Omega}_0 = \tilde{\Omega}_6$). The branch of the function $\mu(\eta)$ for the solutions $\Psi^{1,2,6}_{WKB}(z)$ is fixed by the condition

$$-\pi < \arg \eta < \pi .$$

II. THE SOLUTIONS AT THE NEIGHBOURHOOD OF THE TURNING POINT z_+

Let us fix $\varepsilon > 0$ and consider the following domain near the turning point $z_+ = \frac{1}{2}$:

$$\mathcal{D}_+^\varepsilon = \{ z : |z - \tfrac{1}{2}| \leq \tau^{-1/2 + \varepsilon} \} .$$

Assuming $z \in \mathcal{D}_+^\varepsilon$ we will put in eq. (5.3) the Taylor series expansion of matric $A_0(z,\tau)$ near $z_+ = \frac{1}{2}$, cancelling terms quadratic in $z - \frac{1}{2}$ or $\tau^{-\frac{1}{2}}$. As a result we'll get an equation

$$\frac{d\Psi_0}{dz} = \tau \cdot \left[-4i(z - \tfrac{1}{2})\sigma_3 - 2u \cdot (-x)^{-1/2} \sigma_2 - \frac{2w}{x} \sigma_1 \right] \Psi_0 . \qquad (5.12)$$

Introducing the new variables

$$\zeta = 2\sqrt{2\tau}\,(z - \tfrac{1}{2})\, e^{-\frac{i\pi}{4}} ,$$

$$v = \sqrt{2}\,(-x)^{3/4} e^{\frac{i\pi}{4}} \left(\tfrac{i}{2} u (-x)^{-1/2} + \tfrac{w}{2} x^{-1} \right) , \qquad (5.13)$$

$$v^* = -\sqrt{2}\,(-x)^{3/4} e^{\frac{i\pi}{4}} \left(\tfrac{i}{2} u (-x)^{-1/2} - \tfrac{w}{2} x^{-1} \right) ,$$

we simplify eq. (5.12) to the form

$$\frac{d\Psi_0}{d\zeta} = \begin{pmatrix} \frac{\zeta}{2} & v \\ v^* & -\frac{\zeta}{2} \end{pmatrix} \Psi_0 . \qquad (5.14)$$

Eq. (5.14) may be solved in terms of parabolic cylinder functions. To check this let us differentrate eq. (5.14) by ζ

$$\frac{d^2\Psi_0}{d\zeta^2} = \left[\begin{pmatrix} \frac{1}{2} & 0 \\ 0 & -\frac{1}{2} \end{pmatrix} + \begin{pmatrix} \frac{\zeta}{2} & v \\ v^* & -\frac{\zeta}{2} \end{pmatrix}^2\right]\Psi_0 =$$

$$= \begin{pmatrix} \frac{\zeta^2}{4} + vv^* + \frac{1}{2} & 0 \\ 0 & \frac{\zeta^2}{4} + vv^* - \frac{1}{2} \end{pmatrix}\Psi_0 .$$

It is clear now that elements of the first line of Ψ_0 may be expressed through Weber-Hermite functions $D_\nu(\zeta)$ or $D_{-1-\nu}(i\zeta)$ with the simbol

$$\nu = -vv^* - 1 = 2i\tau\left(\frac{u^2}{4x} - \frac{w^2}{4x^2}\right) - 1 . \tag{5.15}$$

The elements of the second line of Ψ_0 are constructed now directly from eq. (5.14).

We get thus an exact formula for the solution of eq. (5.3) in the small domain D_+^ε near the turning point $z_+ = \frac{1}{2}$:

$$\Psi_0(z) = \begin{pmatrix} D_{-1-\nu}(i\zeta) & D_\nu(\zeta) \\ \dot{D}_{-1-\nu}(i\zeta) & \dot{D}_\nu(\zeta) \end{pmatrix} , \tag{5.16}$$

where „ $\cdot$ " means a transform

$$\psi(\zeta) \mapsto \dot{\psi}(\zeta) = \left(\psi_\zeta - \frac{\zeta}{2}\psi\right)/v , \tag{5.17}$$

and variables ζ , v , ν are defined by eqs. (5.13), (5.15).

Let us discuss more strictly in what sense the formula (5.16) presents an asymptotics for the precise solution of eq. (5.3) in the domain $\mathcal{D}_+^{\varepsilon}$. Designating the cancelled part of matrix $A_0(z,\tau)$ through $R_+(z)$ we have

$$R_+(z) = O(\tau^{-1+2\varepsilon}), \quad \tau \to \infty, \quad z \in \mathcal{D}_+ .$$

The precise solution of eq. (5.3) in the domain $\mathcal{D}_+^{\varepsilon}$ may be presented in the form $\Psi(z) = \chi(z)\Psi_0(z)$. Then for the matrix χ an integral equation takes place, which is equivalent to differential equation (5.3) with initial condition $\Psi(\frac{1}{2}) = \Psi_0(\frac{1}{2})$:

$$\chi(z) =$$

$$= I + \tau \int_{1/2}^{z} \Psi_0(z)\Psi_0^{-1}(\eta) R_+(\eta)\chi(\eta)\Psi_0(\eta)\Psi_0^{-1}(z)\, d\eta . \tag{5.18}$$

This equation is of Volterra type with regular kernel. Thus there exists its unique solution $\chi(z)$ in all the domain $\mathcal{D}_+^{\varepsilon}$ and it is regular function of z. In order to get its asymptotic properties as $\tau \to \infty$ let us remind the condition b) of Theorem 5.1. Assume that it holds for the value υ, then the devision on υ in eq. (5.17) is correct operation from the asymptotic point of view. Furthermore due to the condition a) the parameter ν in eq. (5.16) remains bounded as $\tau \to \infty$. The transform $\Psi_0 \mapsto \Psi_0^{-1}$ does not lead to any singularity in eq. (5.18) because of the well-known formula for the parabolic cylinder functions [53]

$$\det \Psi_0(z) = i \exp\left\{-\frac{\nu\pi i}{2}\right\} \cdot \upsilon^{-1} . \tag{5.19}$$

It is essential that the domain $\mathcal{D}_+^{\varepsilon}$ contains a circular domain

$$\Omega_+^{\varepsilon} = \left\{ z \in \mathcal{D}_+^{\varepsilon} : z - \frac{1}{2} \approx \tau^{-\frac{1}{2}+\varepsilon} \right\} ,$$

where the variable ζ becomes large. Due to the condition c) and the well-known asymptotics for $D_\nu(\zeta)$ and $D_{1-\nu}(i\zeta)$ an estimate for the function $\Psi_0(z)$ is true

$$|\Psi_0(z)| \leqslant C \exp\{|\operatorname{Re} \zeta^2|\}, \ \tau \to \infty . \tag{5.20}$$

This estimate holds, certainly, in all neighbourhood $\mathcal{D}_+^\varepsilon$.

The considerations mentioned above together with estimate (5.20) make it possible to evaluate the kernel of the integral equation (5.18)

$$\int_{1/2}^{z} |\Psi_0(z)\Psi_0^{-1}(\eta)|\cdot|R_+(\eta)|\cdot|\Psi_0(\eta)\Psi_0^{-1}(z)|d\eta =$$

$$= O(\tau^{-3/2+3\varepsilon}), \quad \tau \to \infty, \quad z \in \mathcal{D}_+^\varepsilon ,$$

$$\arg(z-\tfrac{1}{2}) = \arg(\eta - \tfrac{1}{2}) = \frac{k\pi}{2} , \quad k=-1,0,1,2. \tag{5.21}$$

The direct corollary of the estimate (5.21) is an asymptotic evaluation of solution $\overset{\circ}{\Psi}_+(z)$, at the neighbourhood of turning point $z_+ = \frac{1}{2}$. It is formulated in the following lemma.

LEMMA 5.2. There exists a solution of eq. (5.3) $\overset{\circ}{\Psi}_+(z)$ which has the following asymptotic behaviour in the domain $\mathcal{D}_+^\varepsilon$, $\varepsilon < \frac{1}{6}$

$$\overset{\circ}{\Psi}_+(z) = (I + o(1))\Psi_0(z), \quad \tau \to \infty ,$$

$$z \in \mathcal{D}_+^\varepsilon , \quad \arg(z - \tfrac{1}{2}) = \frac{\pi}{2} j , \quad j=-1,0,1,2,$$

where $\Psi_0(z)$ is described by eqs. (5.13), (5.15), (5.16), (5.17).

REMARK 5.1. The proof of Lemma 5.2 goes on if the condition b) is not satisfied for the value ϑ . Then the second alternative is true, i.e. the value ϑ^* is separated from zero. In this case we must change $\Psi_0(z)$ as follows

$$\Psi_0(z) \mapsto \widetilde{\Psi}_0(z) = \begin{pmatrix} \ddot{D}_{-\nu}(i\zeta) & \ddot{D}_{\nu+1}(\zeta) \\ D_{-\nu}(i\zeta) & D_{\nu+1}(\zeta) \end{pmatrix},$$

where „ $\cdot\cdot$ " means the transform $\psi(\zeta) \mapsto \ddot{\psi}(\zeta) = \left(\psi_\zeta + \zeta/2\,\psi\right)/\vartheta^*$.

At the end of the proof it is possible to return to $\Psi_0(z)$, remembering that Ψ_0 and $\widetilde{\Psi}_0$ are linked together by a constant right matrix multiplier as two solutions of the same equation.

REMARK 5.2. More precise reasoning, which we do not produce here, show that lemma 5.2 remains true if the condition b) is not satisfied at all.

III. COMPUTATION OF MATRICES C_K

Due to almost canonical form of WKB-asymptotics (5.11) and assumed way of fixation of solutions Ψ^K_{WKB} with $K=1,2,6$ we have the diagonal matrices $C_K = [\Psi^K_{WKB}]^{-1}\Psi_K$, which could be expressed as follows

$$C_K = \lim_{z\to\infty} \exp\left\{+\tau\int_{1/2}^{z} \mu(\eta,\tau)\,d\eta\,\sigma_3 - \left(\frac{4i}{3}z^3 - iz\right)\sigma_3\right\},$$

$$\arg z = \frac{\pi}{3} j_K, \quad j_K = \begin{cases} K-1, & K=1,2, \\ -1, & K=6 \end{cases}. \tag{5.22}$$

Thus for computation of the leading order in τ of C_K matrices we must get an asymptotics of integral

$$I(z,\tau)=\int_{1/2}^{z}\mu(\eta,\tau)\,d\eta$$

as $\tau\to\infty$, which is uniform for large values of $|z|$.

Let us put $|z|>C$, $\arg z=\frac{\pi}{3}j_K$, fix arbitrary ε , $0<\varepsilon<<\frac{1}{6}$ and split the integral I into a sum

$$I(z,\tau)=I_1(z,\tau)+I_2(z,\tau),$$

$$I_1(z,\tau)=\int_{1/2}^{1/2+\tau^{-1/2+\varepsilon}}\mu(\eta,\tau)d\eta,\quad I_2(z,\tau)=\int_{1/2+\tau^{-1/2+\varepsilon}}^{z}\mu(\eta,\tau)d\eta.$$

Denoting $a^2=-\frac{w^2}{4x^2}+\frac{u^2}{4x}$, we get

$$a^2=O(\tau^{-1}),\quad \frac{u^4}{x^2}=O(\tau^{-2}). \tag{5.23}$$

Then, retaining the leading terms in τ for $\mu(\eta,\tau)$, we obtain an asymptotics for I_1 :

$$I_1=4i\int_{1/2}^{1/2+\tau^{-1/2+\varepsilon}}\sqrt{(\eta^2-1/4)^2+a^2+O(\tau^{-2})}\,d\eta=$$

$$=4i\int_{1/2}^{1/2+\tau^{-1/2+\varepsilon}}\sqrt{(\eta-1/2)^2+a^2+O(\tau^{-3/2+3\varepsilon})}\,d\eta=$$

$$=4i\int_{1/2}^{1/2+\tau^{-1/2+\varepsilon}}\sqrt{(\eta-1/2)^2+a^2}\,d\eta+O(\tau^{-3/2+3\varepsilon})=$$

$$+\,2iy\sqrt{y^2+a^2}\,\Big|_{y=0}^{y=\tau^{-1/2+\varepsilon}}\qquad +$$

$$+ 2ia^2 \ln(y + \sqrt{y^2 + a^2}) \Big|_{y=0}^{y=\tau^{-1/2+\varepsilon}} + O(\tau^{-3/2+3\varepsilon}) .$$

Finally we get

$$I_1 = 2i\tau^{-1+2\varepsilon} + ia^2 - ia^2 \ln a^2 + 2ia^2 \ln \tau^{-1/2+\varepsilon} + 2ia^2 \ln 2 + $$
$$+ O(\tau^{-1-2\varepsilon}) + O(\tau^{-3/2+3\varepsilon}) . \tag{5.24}$$

The second integral I_2 is evaluated similarly:

$$I_2 = 4i \int_{1/2+\tau^{-1/2+\varepsilon}}^{z} (\eta^2 - 1/4)\sqrt{1 + \frac{a^2 + O(\tau^{-2})}{(\eta^2 - 1/4)^2}}\, d\eta =$$

$$= 4i \int_{1/2+\tau^{-1/2+\varepsilon}}^{z} (\eta^2 - 1/4)\, d\eta + 2i(a^2 + O(\tau^{-2})) \int_{1/2+\tau^{-1/2+\varepsilon}}^{z} \frac{d\eta}{\eta^2 - 1/4} =$$

$$= (4/3\, i\eta^3 - i\eta)\Big|_{\eta=1/2+\tau^{-1/2+\varepsilon}}^{\eta=z} + 2i\left[a^2 + O(\tau^{-1-2\varepsilon})\right] \ln \frac{\eta - 1/2}{\eta + 1/2} \Big|_{\eta=1/2+\tau^{-1/2+\varepsilon}}^{\eta=z} .$$

Hence, finally

$$I_2 = \frac{4i}{3} z^3 - iz + \frac{i}{3} + 2ia^2 \ln \frac{z - 1/2}{z + 1/2} - 2i\tau^{-1+2\varepsilon} -$$

$$- 2ia^2 \ln \tau^{-1/2+\varepsilon} + O(\tau^{-3/2+\varepsilon}) + O(\tau^{-1-2\varepsilon}) +$$

$$+ O\left(\tau^{-1-2\varepsilon} \ln \frac{z - 1/2}{z + 1/2}\right) . \tag{5.25}$$

Summing eq. (5.24) and eq. (5.25) we obtain the following asymptotic of integral in question

$$I(z,\tau)=\frac{4i}{3}z^3-iz+\frac{i}{3}+b+2ia^2\ln\frac{z-1/2}{z+1/2}+o(\tau^{-1}),$$

$$\tau\to\infty, \tag{5.26}$$

where $b=ia^2-ia^2\ln a^2+2ia^2\ln 2$.

The asymptotics (5.26) is uniform on z, when $|z|>C$, $\arg z=\frac{\pi}{3}j_k$. Substituting this asymptotics into eq. (5.22), we find that matrices C_k are calculated explicitly in the leading term of τ

$$C_k=\exp\left\{\frac{i\tau}{3}+\tau b\right\}(1+o(1)),$$

$$\tau\to\infty, \quad k=1,2,6. \tag{5.27}$$

It is worth mentioning that C_k appeared to be independent of k. This fact occured due to our agreement (see point I) how to fix the branch of $\mu(\eta)$.

IV. COMPUTATION OF N_k MATRICES

The estimate (5.11) implies that for Ψ^k_{WKB} solution in the domain Ω^{ε}_{+} $(z-1/2\approx\tau^{-1/2+\varepsilon})$ where the matching with $\mathring{\Psi}_{+}$ takes place there is an asymptotics

$$\Psi^k_{WKB}(z)=(I+o(1))\exp\left\{-\tau\sigma_3\int_{1/2}^{z}\mu(\eta,\tau)d\eta\right\},$$

$$\tau\to\infty, \quad z\in\Omega^{\varepsilon}_{+}, \quad \arg(z-1/2)=\frac{\pi}{2}j_k.$$

That is why the basic computational moment at this stage turns out to be the asymptotic expansion of integral $I(z,\tau)$ as $\tau\to\infty$ and z belongs to Ω^{ε}_{+}. It is clear however, that in fact we have already found it, when the estimate for the integral I_1 was produced in eq. (5.24). Thus, assuming as usual $0<\varepsilon<1/6$ we may write down at once

$$\int_{1/2}^{z} \mu(\eta,\tau)\,d\eta = 2i\left(z-\frac{1}{2}\right)^{2} + 2ia^{2}\ln\left(z-\frac{1}{2}\right) + b + o(\tau^{-1}) ,$$

$$\tau\to\infty , \quad z\in\Omega_{+}^{\varepsilon} , \quad \arg\left(z-\frac{1}{2}\right)=\frac{\pi}{2}j_{k} .$$

Introducing the variables (5.13) in the *rhs* of this formula, we immediately get the following expression for WKB-solutions Ψ^{k}_{WKB}, $k=1,2,6$

$$\Psi^{k}_{WKB} = \left(I+o(1)\right)\exp\left\{\sigma_{3}\left(\frac{\zeta^{2}}{4}-(\nu+1)\ln\zeta\right)\right\}N_{0} ,$$

$$\tau\to\infty , \quad \tau\in\Omega_{+}^{\varepsilon} , \quad \arg\left(z-\frac{1}{2}\right)=\frac{\pi}{2}j_{k} , \tag{5.28}$$

where

$$N_{0} = \exp\left\{\sigma_{3}\left(\frac{1}{2}(\nu+1)\ln 8\tau - i(\nu+1)\frac{\pi}{4} - \tau b\right)\right\} .$$

Let us look now at the solution $\mathring{\Psi}_{+}(z)$. Taking the same z as in (5.28), we can derive its asymptotics from the well-known asymptotic expansions for the parabolic cylinder functions. Skipping some trivial calculations we'll present the final result

$$\mathring{\Psi}_{+}(z) = \left(I+o(1)\right)\exp\left\{\sigma_{3}\left(\frac{\zeta^{2}}{4}-(1+\nu)\ln\zeta\right)\right\}\cdot$$

$$\cdot\begin{pmatrix} e^{-(1+\nu)i\frac{\pi}{2}} & 0 \\ 0 & -\frac{1}{v} \end{pmatrix} , \quad \tau\to\infty , \tag{5.29}$$

where $z \in \overset{\varepsilon}{\Omega}_+$, $\arg(z-\frac{1}{2}) = 0$;

$$\overset{\circ}{\Psi}_+(z) = (I + o(1)) \exp\{\sigma_3(\frac{\zeta^2}{4} - (1+\nu)\ln\zeta)\} \cdot$$

$$\begin{pmatrix} e^{-(1+\nu)\frac{i\pi}{2}} & 0 \\ -\frac{i\sqrt{2\pi}\, e^{-\frac{\nu+1}{2}\pi i}}{\upsilon\Gamma(1+\nu)} & -\frac{1}{\upsilon} \end{pmatrix} \quad \tau \to \infty \ , \tag{5.30}$$

where $z \in \Omega_+^{\varepsilon}$, $\arg(z-\frac{1}{2}) = +\frac{\pi}{2}$;

$$\overset{\circ}{\Psi}_+(z) = (I + o(1)) \exp\{\sigma_3(\frac{\zeta^2}{4} - (1+\nu)\ln\zeta)\} \cdot$$

$$\begin{pmatrix} e^{-(1+\nu)\frac{i\pi}{2}} & -\frac{\sqrt{2\pi}\, e^{-\nu\pi i}}{\Gamma(-\nu)} \\ 0 & -\frac{1}{\upsilon} \end{pmatrix}, \quad \tau \to \infty \ , \tag{5.31}$$

where $z \in \Omega_+^{\varepsilon}$, $\arg(z - 1/2) = -\pi/2$.

The leading order in τ of matrices $N_k = \overset{\circ}{\Psi}{}_+^{-1} \Psi_{WKB}^{k}$ may be easily derived now from comparison of eq. (5.28) with eqs.(5.29)-(5.31). The result is

$$N_1 = \begin{pmatrix} e^{(1+\nu)\frac{\pi i}{2}} & 0 \\ 0 & -\upsilon \end{pmatrix} (I + o(1))\, N_0 \ ,$$

$$N_2 = \begin{pmatrix} e^{(1+\nu)\frac{\pi i}{2}} & 0 \\ -\frac{i\sqrt{2\pi}}{\Gamma(1+\nu)} & -\upsilon \end{pmatrix} (I + o(1))\, N_0 \ , \tag{5.32}$$

$$N_6 = \begin{pmatrix} e^{(1+\nu)\frac{\pi i}{2}} & \frac{v\sqrt{2\pi}}{\Gamma(-\nu)} e^{-(1+\nu)\frac{\pi i}{2}} \\ 0 & -v \end{pmatrix} (I + o(1)) N_0 .$$

V. COMPUTATION OF STOKES MATRICES S_1 AND S_6

The substitution of asymptotics (5.32) and (5.27) into eqs. (5.6), (5.7), defining matrices S_1, S_6, leads to the equations

$$S_1 = \begin{pmatrix} 1 & 0 \\ \frac{i\sqrt{2\pi}}{v\Gamma(1+\nu)} \exp\{\frac{2i}{3}\tau + (1+\nu)\ln 8\tau - i(1+\nu)\frac{\pi}{2}\} & 1 \end{pmatrix} + o(1),$$

$$\tau \to \infty ,$$

$$S_6 = \begin{pmatrix} 1 & -\frac{v\sqrt{2\pi}}{\Gamma(-\nu)} \exp\{-\frac{2i}{3}\tau - (\nu+1)\ln 8\tau - i(1+\nu)\frac{\pi}{2}\} \\ 0 & 1 \end{pmatrix} + o(1),$$

$$\tau \to \infty .$$

And here, strictly speaking, is an end of the computations of matrices in question. The only thing to do is to return to the parameters p, q which are the non-diagonal non-zero elements of S_1, S_6. Putting $\delta = -i(1+\nu)$ and transforming Γ-function as

$$\Gamma(-\nu) = \Gamma(1-\nu-1) = -(1+\nu)\Gamma(-1-\nu) = v v^* \Gamma(-i\delta) ,$$

we arrive to eq. (5.2) in the Theorem 5.1, which expresses the leading term of monodromy data for the system (1.9). This accomplishes the proof of Theorem 5.1.

Chapter 6. ASYMPTOTIC SOLUTION TO A DIRECT PROBLEM OF THE MONODROMY THEORY FOR THE SYSTEM (1.26)

The purpose of this Chapter is to prove the theorem analogues to the theorem 5.1, but for the system (1.26).

THEOREM 6.1. Let the parameters u, w of the system (1.26) depend on x in such a way that

a)
$$u = O(x^{-1/2}) + 2k\pi, \quad k \in \mathbb{Z} ,$$
$$w = O(x^{-1/2}) \quad , \quad x \to +\infty , \tag{6.1}$$

b) at least one of the values v, v^*

$$v = -\frac{\sqrt{x}}{4}(u_k + iw)e^{\frac{i\pi}{4}} ,$$

$$v^* = \frac{\sqrt{x}}{4}(u_k - iw)e^{\frac{i\pi}{4}} , \quad u_k = u - 2\pi k ,$$

is separated from zero as $x \to +\infty$,

c)
$$-1 \leq \operatorname{Re} vv^* \leq 0 .$$

Then the monodromy data p, q have the following asymptotic expansions

$$p = -\frac{1}{v^*}\frac{\sqrt{2\pi}}{\Gamma(-i\delta)}\exp\left\{ix - i\delta\ln 4x + \delta\frac{\pi}{2}\right\} + o(1) ,$$
$$q = \frac{i}{v}\frac{\sqrt{2\pi}}{\Gamma(i\delta)}\exp\left\{-ix + i\delta\ln 4x + \delta\frac{\pi}{2}\right\} + o(1) , \tag{6.2}$$

$$i\delta = -vv^* , \quad x \to +\infty .$$

The proof of the theorem 6.1 closely follows the scheme of that of the theorem 5.1. Thus we'll omit much of calculations which repeat the corresponding parts of Chapter 5.

In order to apply WKB method we transform the initial system (1.26) scaling the variable λ :

$$\frac{d\Psi}{dz} = x A_0(z, x)\Psi ,$$
$$z = \lambda x , \tag{6.3}$$
$$A_0 = \left(-\frac{i}{16} + \frac{i}{z^2}\cos u\right)\sigma_3 - \frac{iw}{4z}\sigma_1 - \frac{i}{z^2}\sin u\sigma_2 .$$

The WKB-solution of eq. (6.3) has the structure of eq. (5.4) $(\tau \equiv x!)$ with matrices T and Λ , defined by equations

$$\Lambda = -\mu\sigma_3 , \quad \mu = i\sqrt{\left(\frac{1}{16} - \frac{1}{z^2}\right)^2 - \frac{w^2}{16z^2} + \frac{1-\cos u}{8z^2}} ,$$
$$T = \left[\left(\frac{i}{16} - \frac{i}{z^2}\cos u + \mu\right)\sigma_0 - \frac{1}{z^2}\sin u\sigma_1 + \frac{w}{4z}\sigma_2\right]\frac{8z^2}{i(z^2-16)} . \tag{6.4}$$

From eq. (6.4) and from the estimates (6.1) it follows, that

a) the equation (6.3) has four simple turning points, tending in pairs to two double points:

$$z_\pm = \pm 4$$

b) in a domain, where

$$\sqrt{x}\left|z - \frac{16}{z}\right| \to \infty$$

the matrix T has the estimate

$$T = \left[I - \frac{16}{z^2} B_0 \cos\frac{u}{2}\right]\frac{z^2}{z^2-16} + O\left(\frac{z}{\sqrt{x}(z^2-16)}\right) .$$

It means, in particular, that

$$T \to I\ , \quad z \to \infty\ ,$$
$$T \to B_0 \cos\frac{u}{2}\ , \quad z \to 0\ . \tag{6.5}$$

c) the conjugate Stokes lines tend to the real axis or to the circle $\xi^2 + \eta^2 = 16$, $z = \xi + i\eta$ (see figure 6.1)

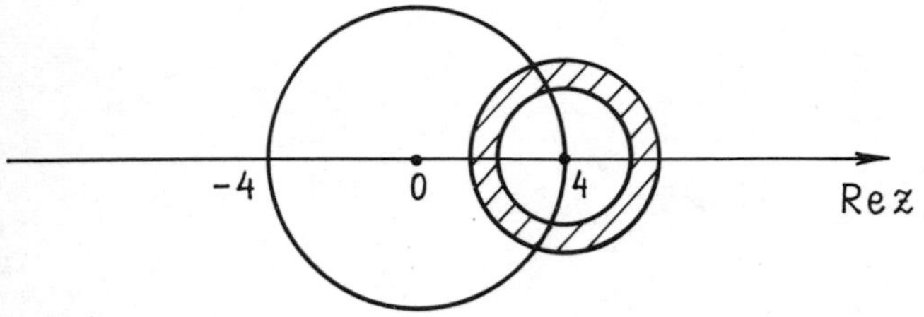

Figure 6.1. Shaded domain - the domain Ω_+^{ε}, where asymptotic matching takes place.

Let us denote through Ω_1 the exterior of the circle $|z| = 4$ without the ray $(-\infty, -4)$, and through $\Omega_\pm$ – the open upper and lower half-planes. Also we'll assume that

$$-\pi < \arg z < \pi\ , \quad z \in \Omega_1 \cup \Omega_+ \cup \Omega_-\ . \tag{6.6}$$

The propositions a) - c) been formulated above, allow us to define at the domains $\Omega_1, \Omega_\pm$ WKB-solutions Ψ^1_{WKB} and $\Psi^\pm_{WKB}$ with double asymptotics of the form

$$\Psi^1_{WKB} = \left[\left(I - \frac{16}{z^2} B_0 \cos\frac{u}{2}\right)\frac{z^2}{z^2-16} + O\left(\frac{z}{\sqrt{x}\,(z^2-16)}\right)\right] \times$$

$$\times \exp\left\{-x\sigma_3 \int_4^z \mu(\eta, x)\, d\eta\right\}\ , \quad \sqrt{x}\,|z-4| \to \infty\ , \tag{6.7}$$

$$\Psi^{\pm}_{WKB} = \left[\left(I - \frac{16}{z^2} B_0 \cos\frac{u}{2} \right) \frac{z^2}{z^2-16} + O\left(\frac{z}{\sqrt{x}(z^2-16)} \right) \right] \times$$

$$\times \exp\left\{ -x\sigma_3 \int_4^z \mu(\eta, x)\, d\eta \right\}, \quad \sqrt{x}\, \left| z - \frac{16}{z} \right| \longrightarrow \infty . \qquad (6.8)$$

The asymptotic expressions (6.5), (6.7), (6.8) show that the solutions Ψ^1_{WKB} and $\Psi^{\pm}_{WKB}$ are linked by right diagonal matrix multipliers with the canonical solutions Ψ^{∞}_1 and $\Psi_{\pm}$ correspondingly. The diagonal matrices in question may be calculated as follows

$$C_1 = \lim_{\substack{z\to\infty \\ \arg z = 0}} \exp\left\{ x\sigma_3 \int_4^z \mu(\eta)\, d\eta - \frac{ixz}{16}\sigma_3 \right\},$$

$$C_1 = \left[\Psi^1_{WKB} \right]^{-1} \Psi^{\infty}_1 , \qquad (6.9)$$

$$C_{\pm} = \lim_{\substack{z\to\infty \\ \arg z = \pm\gamma}} \exp\left\{ x\sigma_3 \int_4^z \mu(\eta)\, d\eta - \frac{ixz}{16}\sigma_3 \right\}, \quad C_{\pm} = \left[\Psi^{\pm}_{WKB} \right]^{-1} \Psi_{\pm}$$

$$\arg z = \pm\gamma , \quad 0 < \gamma < \pi .$$

At the neighbourhood of the turning point

$$\mathcal{D}^{\varepsilon}_{+} = \left\{ z : |z-4| < x^{-1/2+\varepsilon} , \quad 0 < \varepsilon < \frac{1}{6} \right\}$$

the equation (6.2) is simplified to equation of the form of eq. (5.14) with the parameters

$$v = -\frac{\sqrt{x}}{4} (u_K + iw) e^{\frac{i\pi}{4}} ,$$
$$v^* = \frac{\sqrt{x}}{4} (u_K - iw) e^{\frac{i\pi}{4}} \qquad (6.10)$$

and independent variable

$$\zeta = \frac{\sqrt{x}}{4}(z-4)e^{-\frac{i\pi}{4}} . \tag{6.11}$$

Thus the equation (6.3) has a solution Ψ_0 in the domain $\mathcal{D}_+^{\varepsilon}$ with the following asymptotic behaviour

$$\Psi_0(z) = (I + o(1)) \begin{pmatrix} D_{-1-\nu}(i\zeta) & D_\nu(\zeta) \\ \dot{D}_{-1-\nu}(i\zeta) & \dot{D}_\nu(\zeta) \end{pmatrix} , \tag{6.12}$$

where $\tau \to \infty \ , \ z \in \mathcal{D}_+^{\varepsilon} \ , \ \arg(z-4) = 0 \ , \ \pm \frac{\pi}{2} \ ,$

$$\nu = -v v^* - 1 = i \frac{x}{16}(w^2 + u_k^2) - 1 . \tag{6.13}$$

The investigation of integral $\int_4^z \mu(\eta)\, d\eta$ at the neighbourhood of infinity and in circular domain

$$\Omega_+^{\varepsilon} = \{ z : |z-4| \approx x^{-1/2+\varepsilon} , \quad 0 < \varepsilon < 1/6 \}$$

repeats word for word the corresponding part of Chapter 5. As a result we obtain asymptotic formulas

$$\int_4^z \mu(\eta)\, d\eta = \frac{i}{16} z + \frac{i}{z} - \frac{i}{2} + b + i a^2 \ln \frac{z-4}{z+4} + o(\tau^{-1}) ,$$

$$x \to \infty \ , \ |z| > C , \ \arg z = 0 , \ \pm \gamma , \tag{6.14}$$

and

$$\int_4^z \mu(\eta)\, d\eta = \frac{i}{64}(z-4)^2 + i a^2 \ln(z-4) + b - 3 i a^2 \ln 2 + o(\tau^{-1}) ,$$

$$x \to \infty \ , \ z \in \Omega_+^{\varepsilon} \ , \ \arg(z-4) = 0 , \ \pm \frac{\pi}{2} \ , \tag{6.15}$$

where we have denoted

$$a^2 = \frac{w^2 + u^2}{16} \quad , \quad b = \frac{i}{2} a^2 - \frac{i}{2} a^2 \ln 64 a^2 + 4 i a^2 \ln 2 \ .$$

The substitution of asymptotics (6.14) into eq. (6.9) leads to the expression for C_1, $C_\pm$ matrices

$$C_1 = C_\pm = \exp\left\{\left(-\frac{xi}{2} + bx\right)\sigma_3\right\} + o(1) \ . \tag{6.16}$$

On the other hand, from the formulae (6.15), (6.14) together with eqs. (6.7), (6.8) we obtain the leading term in x of the matrices $N_1 = \Psi_0^{-1} \Psi^1_{WKB}$ and $N_\pm = \Psi_0^{-1} \Psi^\pm_{WKB}$

$$N_1 = \begin{pmatrix} e^{(1+\nu)\frac{i\pi}{2}} & 0 \\ 0 & -v \end{pmatrix} (I + o(1)) \exp\left\{\left(\frac{\nu+1}{2} \ln 4x - \frac{\nu+1}{4} i\pi - bx\right)\sigma_3\right\},$$

$$N_+ = \begin{pmatrix} e^{(1+\nu)\frac{i\pi}{2}} & 0 \\ -\frac{i\sqrt{2\pi}}{\Gamma(1+\nu)} & -v \end{pmatrix} (I + o(1)) \exp\left\{\left(\frac{\nu+1}{2} \ln 4x - \frac{\nu+1}{4} i\pi - bx\right)\sigma_3\right\},$$

(6.17)

$$N_- = \begin{pmatrix} e^{(1+\nu)\frac{i\pi}{2}} & \frac{v}{\Gamma(-\nu)} \sqrt{2\pi}\, e^{-\frac{i\pi}{2}(1+\nu)} \\ 0 & -v \end{pmatrix} (I + o(1)) \cdot$$

$$\cdot \exp\left\{\left(\frac{\nu+1}{2} \ln 4x - i \frac{\nu+1}{2} \pi - bx\right)\sigma_3\right\} \ .$$

The monodromy data for the system (1.26) are reconstructed through the Stokes matrices

$$S_+^{(\infty)} = [\Psi_1^\infty]^{-1} \Psi_+ = C_1^{-1} N_1^{-1} N_+ C_+ \quad ,$$

$$S_-^{(\infty)} = [\Psi_1^\infty]^{-1} \Psi_- = C_1^{-1} N_1^{-1} N_- C_+ \quad .$$

Equations (6.16), (6.17) produce the asymptotic expressions for $S_\pm^{(\infty)}$ matrices:

$$S_+^\infty = \begin{pmatrix} 1 & 0 \\ \frac{i}{\vartheta} \frac{\sqrt{2\pi}}{\Gamma(1+\nu)} \exp\{-ix+(1+\nu)\ln 4x - i\frac{\nu+1}{2}\pi\} & 1 \end{pmatrix} + o(1) \quad ,$$

$$S_-^\infty = \begin{pmatrix} 1 & \frac{\vartheta}{\Gamma(-\nu)}\sqrt{2\pi}\, \exp\{ix-(1+\nu)\ln 4x - i\frac{\nu+1}{2}\pi\} \\ 0 & 1 \end{pmatrix} + o(1),$$

$$x \to +\infty .$$

The theorem 6.1 is proved completely.

Chapter 7. THE MANIFOLD OF SOLUTIONS OF PAINLEVÉ II EQUATION DECREASING AS $x \to -\infty$. PARAMETRIZATION OF THEIR ASYMPTOTICS THROUGH THE MONODROMY DATA. ABLOWITZ-SEGUR CONNECTION FORMULAE FOR REAL-VALUED SOLUTIONS DECREASING EXPONENTIALLY AS $x \to +\infty$

We begin this Chapter with a simple remark.

PROPOSITION 7.1. The isomonodromic deformation condition for coefficients of eq. (1.9) is matched with estimates (5.1) for corresponding solution of Painlevé equation (3.3) if

$$p \cdot q < 1 \quad . \tag{7.1}$$

PROOF. Multiplying term by term both equalities (5.2) we get

$$p \cdot q = -\frac{2\pi e^{\pi\delta}}{\delta\,\Gamma(i\delta)\Gamma(-i\delta)} = 2ie^{\pi\delta}\sin(i\delta\pi) + o(1)\ .$$

It yields that

$$i v v^* = \delta = \frac{1}{2\pi}\ln(1-pq) + o(1). \tag{7.2}$$

Suppose that inequality (7.1) does not hold. It means that δ has nonvanishing imaginary part as $x \to -\infty$. Then, according to the assumed p and q independence of x, we conclude from the identities (5.2) that either v or v^* have a leading term $(-x)^{3/2|\mathrm{Im}\,\delta|}$ as $x \to -\infty$. This fact contradicts with estimates (5.1), in assumption of which the formulae (5.2) were derived. On the other hand if (7.1) holds then δ has to be real-valued due to (7.2). Moreover, from eq. (7.2) follows also its boundness. Since δ is bounded the estimates (5.1) are obtained directly from the formulae (5.2). The proposition is proved.

On other words the parametrization of solutions of Painlevé II equation by the monodromy data for eq. (1.9) induce the imbedding of the set of decreasing as $x \to -\infty$ solutions into the manifold

$$\mathbb{M} = \{(p, q) \in \mathbb{C}^2 : pq < 1\} . \tag{7.3}$$

The fact that the image of this embedding is nonempty follows from the easily verified statement about the existence of solutions of PII equation with asymptotics (5.1). It would be shown below that this image coinside with the whole $\mathbb{M}$ (see Remark 7.3).

The manifold $\mathbb{M}$ has real dimension equal to 3 and it is naturally decomposed into a sum of four components:

$$\mathbb{M} = \mathbb{M}_r \cup \mathbb{M}_q \cup \mathbb{M}_p \cup \mathbb{M}_i \ , \tag{7.4}$$

where

$$\mathbb{M}_r = \{(p, q) \in \mathbb{C}^2 : \arg q = -\arg p,\ 0 < |pq| < 1\} ,$$

$$\mathbb{M}_i = \{(p, q) \in \mathbb{C}^2 : \arg q = -\arg p + \pi,\ p, q \neq 0\}$$

are the submanifolds of full dimension, and

$$\mathbb{M}_p = \{(p, q) \in \mathbb{C}^2 : q = 0\} ,$$

$$\mathbb{M}_q = \{(p, q) \in \mathbb{C}^2 : p = 0\}$$

are two-dimensional "separatrice" submanifolds. For every component of this decomposition it is possible to give an exact description in terms of p, q of corresponding leading term of asymptotics of u, w as $x \to -\infty$. In order to write out this formulae it is necessary only to resolve the equalities (5.2) with respect to u and

w taking into account eq. (7.2). We'll omit trivial algebraic calculations and formulate the final result

THEOREM 7.1. Let the monodromy data $(p,q)\in M$ correspond to decreasing as $x\to-\infty$ solution of PII equation. Then according to decomposition (7.4) the leading term of asymptotics $u(x)$ is given by the following formulae:

a) $(p,q)\in M_r$:

$$u(x)=\frac{\alpha}{\sqrt{2}}(-x)^{-1/4}\Big[(\beta+\frac{1}{\beta})\cos\{\frac{2}{3}(-x)^{3/2}-\frac{3}{2}\alpha^2\ln(-x)+\varphi\}+$$

$$+i(\beta-\frac{1}{\beta})\sin\{\frac{2}{3}(-x)^{3/2}-\frac{3}{2}\alpha^2\ln(-x)+\varphi\}\Big]+o(1),$$

$$x\to-\infty, \tag{7.5}$$

$$\alpha^2=-\frac{1}{2\pi}\ln(1-|pq|),\quad \beta^2=\left|\frac{q}{p}\right|,\quad \alpha,\beta>0,$$

$$\varphi=-3\alpha^2\ln 2-\frac{\pi}{4}-\arg\Gamma(-i\alpha^2)-\arg p.$$

b) $(p,q)\in M_i$:

$$u(x)=\frac{\alpha}{\sqrt{2}}(-x)^{-1/4}\Big[(\beta-\frac{1}{\beta})\cos\{\frac{2}{3}(-x)^{3/2}+\frac{3}{2}\alpha^2\ln(-x)+\varphi\}+$$

$$+i(\beta+\frac{1}{\beta})\sin\{\frac{2}{3}(-x)^{3/2}+\frac{3}{2}\alpha^2\ln(-x)+\varphi\}\Big]+o(1), \tag{7.6}$$

$$x\to-\infty,$$

$$\alpha^2=\frac{1}{2\pi}\ln(1+|pq|),\quad \beta^2=\left|\frac{q}{p}\right|;\quad \alpha,\beta>0,$$

$$\varphi=3\alpha^2\ln 2-\frac{\pi}{4}-\arg\Gamma(i\alpha^2)-\arg p.$$

c) $(p, q) \in \mathbb{M}_p$:

$$u(x) = \frac{p}{2\sqrt{\pi}} (-x)^{-1/4} \exp\left\{-i\frac{2}{3}(-x)^{3/2} + \frac{3i\pi}{4}\right\} + o(1) , \quad x \to -\infty , \tag{7.7}$$

d) $(p, q) \in \mathbb{M}_q$:

$$u(x) = \frac{q}{2\sqrt{\pi}} (-x)^{-1/4} \exp\left\{i\frac{2}{3}(-x)^{3/2} - \frac{3i\pi}{4}\right\} + o(1) , \quad x \to -\infty . \tag{7.8}$$

Moreover, the asymptotics (7.5) - (7.8) admit differentiation in x.

REMARK 7.1. The last statement is a consequence of explicit form of asymptotics for w function, derived from eq. (5.2), and the fact that $w = u_x$ under the isomonodromic condition.

REMARK 7.2. It is obvious that

$$\mathbb{M}_p \cup \mathbb{M}_q = \partial\mathbb{M}_r = \partial\mathbb{M}_i .$$

In terms of solutions of PII equation this geometric fact finds its explanation in the structure of asymptotic formulae. Namely, the formulae (7.5), (7.6) are transformed into (7.7) when $\beta \to 0$ and into (7.8) when $\beta \to \infty$.

REMARK 7.3. The existence of solutions of PII equation with asymptotics (7.5) - (7.8) where α, β, φ are arbitrary is a well-known fact ([33]) which can be established independently. The formulae (7.5) - (7.8) may be explicitely resolved in a trivial way with respect to α, β and φ . Thus we may claim that if one takes any $(p, q) \in \mathbb{M}$ then there exists a solution of PII equation with asymptotics as $x \to -\infty$ given by one of the formulae (7.5) - (7.8).

REMARK 7.4. The manifold $\mathbb{M}_r$ contains the submanifold $\mathbb{M}_{\mathbb{R}}$ of all real-valued decreasing as $x \to -\infty$ solutions of PII equation:

$$\mathbb{M}_{\mathbb{R}} = \{(p,q) \in \mathbb{M}_{r} : p = \bar{q}\} .$$

The asymptotics (7.5) for $u(x) \in \mathbb{M}_{\mathbb{R}}$ becomes more simple

$$u(x) = \sqrt{2}\,\alpha(-x)^{-1/4} \cos\left\{\frac{2}{3}(-x)^{3/2} - \frac{3}{2}\alpha^2 \ln(-x) + \varphi\right\} + o(1),$$

$$x \to -\infty \quad , \tag{7.9}$$

$$\alpha > 0 , \quad \alpha^2 = -\frac{1}{2\pi} \ln(1 - |p|^2) ,$$

$$\varphi = -3\alpha^2 \ln 2 - \frac{\pi}{4} - \arg\Gamma(-i\alpha^2) - \arg p .$$

In a similar way the manifold $\mathbb{M}_{i}$ contains the submanifold $\mathbb{M}_{I}$ of all pure imaginary decreasing as $x \to -\infty$ solutions of PII equation

$$\mathbb{M}_{I} = \{(p,q) \in \mathbb{M}_{r} : p = -\bar{q}\} .$$

The corresponding asymptotics takes the form

$$u(x) = i\sqrt{2}\,\alpha(-x)^{-1/4} \sin\left\{\frac{2}{3}(-x)^{3/2} + \frac{3}{2}\alpha^2 \ln(-x) + \varphi\right\} + o(1) ,$$

$$x \to -\infty , \tag{7.10}$$

$$\alpha > 0 , \quad \alpha^2 = \frac{1}{2\pi} \ln(1 + |p|^2) ,$$

$$\varphi = 3\alpha^2 \ln 2 - \frac{\pi}{4} - \arg\Gamma(i\alpha^2) - \arg p .$$

Let us introduce now an auxiliary one-dimensional submanifold of the manifold $\mathbb{M}_{\mathbb{R}}$:

$$\mathring{\mathbb{M}}_{\mathbb{R}} = \{(p,q) \in \mathbb{M}_{\mathbb{R}} , p \in i\mathbb{R}\} .$$

As it follows from the theorem 4.2 in Chapter 4 there is a possibility to describe in terms of $p, q \in \overset{\circ}{\mathbb{M}}_{\mathbb{R}}$ the asymptotic behaviour of solutions as $x \to +\infty$. Thus, bringing together formulae (7.9) and (4.9) we obtain the one of the most interesting results of modern theory of Painlevé equations - the connection formulae for one-parameter set of PII solutions, decreasing exponentially at $-\infty$ and oscillating at $+\infty$.

THEOREM 7.2. (Ablowitz-Segur). For any $a : -1 < a < 1$, $a \neq 0$ PII equation has a smooth solution $u(x; a)$ with the following asymptotic behaviour:

$$u(x) = \frac{a}{2\sqrt{\pi}} x^{-1/4} e^{-2/3 x^{3/2}} (1 + o(1)), \quad x \to +\infty,$$

$$u(x) = \sqrt{2}\, d (-x)^{-1/4} \cos\left\{ \frac{2}{3}(-x)^{3/2} - \frac{3}{2} d^2 \ln(-x) + \varphi \right\} (1 + o(1)), \tag{7.11}$$

$$x \to -\infty.$$

The connection between asymptotic parameters (d, φ) and a are given by the formulae

$$d^2 = -\frac{1}{2\pi} \ln(1 - a^2),$$

$$\varphi = -3 d^2 \ln 2 - \arg \Gamma(-i d^2) + \frac{\pi}{2} \operatorname{sign} a - \frac{\pi}{4}. \tag{7.12}$$

Clearly, in the case of purely imaginary solutions there exists analogues submanifold:

$$\overset{\circ}{\mathbb{M}}_I = \{ (p, q) \in \mathbb{M}_I : p \in \mathbb{R} \}.$$

The asymptotic formulae, associated with it are obtained from (7.11) and (7.12) by a formal transform

$$a \mapsto ia, \quad d \mapsto id, \quad \cos \mapsto \sin,$$

$$\frac{\pi}{2} \operatorname{sign} a - \frac{\pi}{4} \mapsto -\frac{\pi}{2} \operatorname{sign} a - \frac{\pi}{4},$$

and by omitting the restriction $|a| < 1$. However, in the purely imaginary case there is essential difference from the real-valued situation. We'll show in Chapter 9 that it is possible to describe explicitly the asymptotic behaviour at $+\infty$ of solutions from $\mathbb{M}_I/\mathring{\mathbb{M}}_I$ i.e. general two-parameter solutions. The lack of symmetry between real and imaginary cases is explained quite simply. The reason is that every imaginary solution of PII equation has to be smooth, whereas the real-valued solution might have poles. As a matter of fact it is well-known (see [3]) that the solutions to PII equation might have singularities of the pole type. On the other hand, a behaviour $u(x) \sim ic/(x-x_0)^n$, $c \in \mathbb{R}$, is prohibited by the sign before the nonlinearity. It is quite natural to suppose that real-valued solutions with singularities tending to $+\infty$ in x might be exactly the solutions from $\mathbb{M}_{\mathbb{R}}/\mathring{\mathbb{M}}_{\mathbb{R}}$. Thus instead of the problem of calculation of asymptotic parameters through the monodromy data a new problem may be posed. It consists of a description in terms of p and q of asymptotic distribution of singularities as $x \to +\infty$. In Chapters 10, 11 we'll show how the similar problem may be treated in the case of Painlevé III and II equations.

Chapter 8. THE MANIFOLD OF SOLUTIONS TO PAINLEVÉ III EQUATION. THE CONNECTION FORMULAE FOR THE ASYMPTOTICS OF REAL-VALUED SOLUTIONS TO THE CAUCHY PROBLEM

The very same reasoning as at the beginning of Chapter 7 (see the proof of the Proposition 7.1 together with formulae (6.2)) leads us to the conclusion, that the decreasing (modulo 2π) solutions of PIII equation

$$u_{xx} + \frac{1}{x} u_x + \sin u = 0 \tag{8.1}$$

are parametrized by the points of manifold

$$\mathbb{M} = \{ (p,q) \in \mathbb{C}^2 : pq > -1 \} . \tag{8.2}$$

The decomposition similar to that of (7.4) takes place:

$$\mathbb{M} = \mathbb{M}_r \cup \mathbb{M}_q \cup \mathbb{M}_p \cup \mathbb{M}_i , \tag{8.3}$$

where

$$\mathbb{M}_r = \{ (p,q) \in \mathbb{C}^2 : \arg q = -\arg p ,\ p,q \neq 0 \} ,$$

$$\mathbb{M}_i = \{ (p,q) \in \mathbb{C}^2 : \arg q = -\arg p + \pi ,\ 0 < |pq| < 1 \} ,$$

$$\mathbb{M}_p = \{ (p,q) \in \mathbb{C}^2 : q = 0 \} ,\quad \mathbb{M}_q = \{ (p,q) \in \mathbb{C}^2 : p = 0 \} .$$

THEOREM 8.1. Let the monodromy data $(p,q) \in \mathbb{M}$ correspond to the decreasing ($\mathrm{mod}\, 2\pi$) as $x \to +\infty$ solution $u(x)$ of PIII equation. Then, according to the decomposition (8.3), the leading term of asymptotics of $u(x)$ is given by the following formulae:

a) $(p,q) \in \mathbb{M}_r$:

$$u(x)=2\pi K+\frac{\alpha}{2}x^{-1/2}\Big[(\beta+\frac{1}{\beta})\cos\{x-\frac{\alpha^2}{16}\ln x+\varphi\}+i(\beta-\frac{1}{\beta})\cdot$$

$$\cdot\sin\{x-\frac{\alpha^2}{16}\ln x+\varphi\}\Big]+o(1)\,,\quad x\to+\infty\,, \tag{8.4}$$

$$\alpha^2=\frac{8}{\pi}\ln(1+|pq|)\,,\quad \beta^2=\Big|\frac{q}{p}\Big|\,,\quad \alpha,\beta>0\,,$$

$$\varphi=-\frac{\alpha^2}{8}\ln 2+\frac{3\pi}{4}-\arg\Gamma(-\frac{i\alpha^2}{16})-\arg p\,.$$

b) $(p,q)\in M_i$:

$$u(x)=2\pi K+\frac{\alpha}{2}x^{-1/2}\Big[(\beta-\frac{1}{\beta})\cos\{x+\frac{\alpha^2}{16}\ln x+\varphi\}+$$

$$+\,i(\beta+\frac{1}{\beta})\sin\{x+\frac{\alpha^2}{16}\ln 2+\varphi\}\Big]+o(1)\,,\quad x\to+\infty\,, \tag{8.5}$$

$$\alpha^2=-\frac{8}{\pi}\ln(1-|pq|)\,,\quad \beta^2=\Big|\frac{q}{p}\Big|\,;\quad \alpha,\beta>0\,,$$

$$\varphi=\frac{\alpha^2}{8}\ln 2+\frac{3\pi}{4}-\arg\Gamma(\frac{i\alpha^2}{16})-\arg p\,.$$

c) $(p,q)\in M_p$:

$$u(x)=2\pi K+\alpha\sqrt{\frac{2}{\pi}}\,x^{-1/2}\exp\{-ix-\frac{i\pi}{4}+i\varphi\}+o(1)\,, \tag{8.6}$$

$$x\to+\infty\,,\quad \alpha e^{i\varphi}=p\,,$$

d) $(p,q)\in M_q$:

$$u(x)=2\pi K+\alpha\sqrt{\frac{2}{\pi}}\cdot x^{-1/2}\exp\{ix+\frac{i\pi}{4}+i\varphi\}+o(1)\,,$$

$$x\to+\infty\,,\quad \alpha e^{i\varphi}=q\,. \tag{8.7}$$

The asymptotics (8.4) - (8.7) are differentiable in x .

REMARK 8.1. The system (1.26) is invariant with respect to the transfer $u\mapsto u+2K\pi$. Therefore the integer K in the formulae

(8.4) - (8.7) has not to be reconstructed through the monodromy data p, q. The reason is that the whole class of solutions $u(x) \bmod 2\pi$ has been associated with the same point $(p,q)\in \mathbb{M}$.

REMARK 8.2. The statement similar to that of the Remark 7.3 takes place here, i.e. for any $(p,q)\in \mathbb{M}$ there exists a solution of PIII with the asymptotics given by one of the formulae (8.4) - (8.7).

REMARK 8.3. The submanifold

$$\mathbb{M}_{\mathbb{R}} = \{(p,q)\in \mathbb{M}_r : |p|=|q|\}$$

corresponds to the real-valued solutions of PIII equation. The asymptotics of those solutions has the form

$$u(x) = 2\pi k + \alpha x^{-1/2} \cos\left\{x - \frac{\alpha^2}{16}\ln x + \varphi\right\} + o(1),$$

$$x \to +\infty, \tag{8.8}$$

$$\alpha^2 = \frac{8}{\pi}\ln(1+|p^2|), \quad \alpha > 0,$$

$$\varphi = -\frac{\alpha^2}{8}\ln 2 + \frac{3\pi}{4} - \arg\Gamma\left(-\frac{i\alpha^2}{16}\right) - \arg p.$$

Similarly the pure imaginary $(\bmod 2\pi)$ solutions decreasing $(\bmod 2\pi)$ as $x \to +\infty$ are parametrized by the submanifold

$$\mathbb{M}_I = \{(p,q)\in \mathbb{M}_i : |p|=|q|\}.$$

The corresponding asymptotic formulae is written as follows

$$u(x) = 2\pi k + \alpha\cdot x^{-1/2}\, i \sin\left\{x + \frac{\alpha^2}{16}\ln x + \varphi\right\} + o(1),$$

$$x \to +\infty, \tag{8.9}$$

$$\alpha^2 = -\frac{8}{\pi}\ln(1-|p|^2), \quad \alpha > 0,$$

$$\varphi = \frac{\alpha^2}{8}\ln 2 + \frac{3\pi}{4} - \arg\Gamma\left(\frac{i\alpha^2}{16}\right) - \arg p.$$

We are now going to construct explicit connection formulae for the solutions of PIII equation (8.1) analogues to those of Section 7 for the PII equation. However there is an essential difference in the structure of these formulae for PII and PIII cases. As a matter of fact the solutions of PIII equation have two immovable singular points $(x=0\ ,\ x=\infty)$, while the solutions of PII equation have the only one of it $(x=\infty)$. Therefore the correct connection problem in PIII case is posed through the Cauchy problem on the half-line $(0,\infty)$. More precisely, it is necessary to construct the asymptotic solutions as $x\to 0$ and $x\to\infty$ and to link the parameters of these asymptotics so that they would represent the asymptotics of solution of the Cauchy problem.

It is easy to prove that for any $r, s\in\mathbb{C}$ such that $|\mathrm{Im}\, r|<2$, there exists a solution of PIII equation with the following asymptotic behaviour near the origin:

$$u(x)=r\ln x+s+O(x^{2-|\mathrm{Im}\, r|}),\ x\to 0. \tag{8.10}$$

Starting from the Cauchy data r, s it is required to construct the leading term of asymptotics to this solution (if it exists, of course) as $x\to+\infty$. According to our ideology proclaimed in this paper it is necessary to compute the monodromy data associated with the solution (8.10) and to find out the conditions on r and s that provide the solution being contained in the manifold M. After this being done it remains only to apply the formulae (8.4) - (8.7) which give the expressions for the asymptotics in question.

We begin with the crucical point of all this program - the calculation of the monodromy data.

THEOREM 8.2. The monodromy data p, q associated with the solution (8.10) are expressed in the form

$$p = \frac{Ae^{-\pi r/4} - Be^{\pi r/4}}{A+B}, \qquad q = \frac{Be^{-\pi r/4} - Ae^{\pi r/4}}{A+B}, \tag{8.11}$$

where

$$A = 2^{3ir/2} e^{is/2} \Gamma^2\left(\frac{1}{2} + \frac{ir}{4}\right),$$

$$B = 2^{-3ir/2} e^{-is/2} \Gamma^2\left(\frac{1}{2} - \frac{ir}{4}\right).$$

PROOF. The monodromy data for the system (1.26) are reconstructed through the connection matrix (see example 2 in Section 1):

$$Q = \left[\Psi_1^{(0)}\right]^{-1} \Psi_1^{(\infty)}. \tag{8.12}$$

We remaind that under the assumption $w = u_x$, where $u(x)$ is the solution of PIII equation (8.1) the matrix Q is independent of x . Therefore it is sufficient to find out the asymptotic solutions of $\Psi_1^{(0)}(\lambda, x)$ and $\Psi_1^{(\infty)}(\lambda, x)$ as $x \to 0$, assuming $w = u_x$ and $u = u(x)$ of the form (8.10) being placed into the system (1.26). Note that we need the asymptotics of Ψ-functions in λ to be known just on the ray $\arg \lambda = 0$.

Let us begin with $\Psi_1^{(\infty)}$ and seek its asymptotics in the form

$$\Psi_1^{(\infty)}(\lambda, x) = \psi(\xi, xu_x)\left[1 + x^{2-|\mathrm{Im}\, r|}\Phi(\xi, x)\right], \tag{8.13}$$

where $\xi = x^2\lambda/16$ and the ψ matrix satisfies the "abridged" system (1.26):

$$\frac{d\psi}{d\xi} = \left\{-i\sigma_3 - \frac{ixu_x}{4\xi}\sigma_1\right\}\psi. \tag{8.14}$$

The solution of eq. (8.14) is calibrated by the condition (1.41)

$$\Psi \to \exp\{-i\sigma_3 \xi\}, \ \xi \to \infty, \ \arg \xi = 0. \tag{8.15}$$

For the components of Ψ matrix we obtain the Bessel equations following directly from eq. (8.14)

$$\frac{d^2}{d\xi^2} v_1 + \left(1 + \frac{a_+}{\xi^2}\right) v_1 = 0, \tag{8.16}$$

$$\frac{d^2}{d\xi^2} v_2 + \left(1 + \frac{a_-}{\xi^2}\right) v_2 = 0,$$

where

$$v_1 = \Psi_{11} + \Psi_{21}, \quad v_2 = \Psi_{11} - \Psi_{21}, \quad a_\pm = \frac{x^2 u_x^2}{16} \pm i \frac{x u_x}{4}.$$

Thus we have $v_{1,2} = \sqrt{\xi}\, Z_{\nu_{1,2}}(\xi)$, where $Z_{\nu_{1,2}}(\xi)$ are the Bessel functions with the symbols

$$\nu_{1,2} = \frac{1}{2} \pm \frac{i x u_x}{4}.$$

Taking into account the calibration (8.15) and the well-known asymptotics of the Hankel function we conclude that

$$\Psi(\xi) =$$

$$= \sqrt{\frac{\pi}{2}}\, e^{-\pi x u_x} \sqrt{\xi} \begin{pmatrix} -iH^{(2)}_{\nu}(\xi) + H^{(2)}_{\nu-1}(\xi), & iH^{(1)}_{-\nu+1}(\xi) - H^{(1)}_{-\nu}(\xi) \\ iH^{(2)}_{\nu}(\xi) + H^{(2)}_{\nu-1}(\xi), & iH^{(1)}_{-\nu+1}(\xi) + H^{(1)}_{-\nu}(\xi) \end{pmatrix}, \tag{8.17}$$

where $\nu = \frac{1}{2} + \frac{i x u_x}{4}$.

The substitution of (8.13) into eq. (1.26) and scaling the terms of order $x^{2-|\mathrm{Im}\,\nu|}$ yields the estimate for the residual term Φ in asymptotics (8.13):

$$\Phi = O(\xi^{-1}), \quad \arg \xi = 0. \tag{8.18}$$

On the other hand, making use of the obvious equality $xu_x = = \nu + O(x^{2-|\mathrm{Im}\,\nu|})$ we derive from (8.17)

$$\Psi_1^{(\infty)} = \psi(\xi, \nu)(1 + O(x^{2-|\mathrm{Im}\,\nu|})),$$

where $\psi(\xi, \nu)$ is defined by the formula (8.17) where xu_x has to be replaced by ν. Bringing together the latter estimate with that of (8.18) we obtain finally

$$\Psi_1^{(\infty)} = \psi(\xi,\nu)\{1 + O[x^{2-|\mathrm{Im}\,\nu|}(1+\xi^{-1})]\}. \tag{8.19}$$

Let us pass now to the construction of $\Psi_1^{(0)}$ matrix. It is convenient to make the change of variables

$$\lambda = \zeta^{-1}$$

in eq. (1.26) and seek the solution in the form

$$\Psi_1^{(0)} = \varphi(\zeta, xu_x, u)(1 + x^2\chi(\zeta, x)). \tag{8.20}$$

Here the matrix φ satisfies the "abridged" system (1.26)

$$\frac{d\varphi}{d\zeta} = \left\{-i\sigma_3 \cos u + i\sigma_2 \sin u + \frac{ixu_x}{4\zeta}\sigma_1\right\}\varphi, \tag{8.21}$$

calibrated by the condition (1.41)

$$\varphi \to (\cos\frac{u}{2} - i\sigma_1 \sin\frac{u}{2})\exp(-i\sigma_3\zeta),$$
$$\zeta \to \infty, \quad \arg\zeta = 0. \tag{8.22}$$

It is easy to check in the same way as above, that eq. (8.21) is explicitly solved through the Bessel functions. Making use of the calibration condition (8.22), we have

$$\varphi(\zeta, xu_x, u) = \frac{1}{2}\sqrt{\frac{\pi}{2}}\, e^{-\frac{\pi x u_x}{8}} \zeta^{1/2} \times$$

$$\times \begin{pmatrix} -ie^{-iu/2} H_\nu^{(2)} + e^{iu/2} H_{\nu-1}^{(2)}, & -ie^{iu/2} H_{-\nu+1}^{(1)} + e^{-iu/2} H_{-\nu}^{(1)} \\ -ie^{-iu/2} H_\nu^{(2)} - e^{iu/2} H_{\nu-1}^{(2)}, & ie^{iu/2} H_{-\nu+1}^{(1)} + e^{-iu/2} H_{-\nu}^{(1)} \end{pmatrix}, \qquad (8.23)$$

where $\nu = \frac{1}{2} + \frac{ixu_x}{4}$, $H_\nu = H_\nu(\zeta)$.

The same reasoning as above in the case of $\Psi_1^{(\infty)}$ yields the estimate

$$\Psi_1^{(0)} = \varphi(\zeta, x, \tau \ln x + s)\left\{1 + O(x^2\zeta^{-1} + x^{2-|\mathrm{Im}\,\tau|}\right\}. \qquad (8.24)$$

Rescaling the estimates (8.19) and (8.24) to the variable λ, we conclude that in the domain

$$x^{-|\mathrm{Im}\,\tau|} \ll \lambda \ll x^{-2} \qquad (8.25)$$

the residual terms in both estimates become small in x simultaneousy. Therefore the matrix Q may be calculated as $x \to 0$ through the following formula

$$Q = \lim_{x\to 0} \varphi^{-1} \Psi\Big|_{\lambda = x^{-2+\varepsilon}}, \quad 0 < \varepsilon < 2 - |\mathrm{Im}\,\tau|. \qquad (8.26)$$

At the point $\lambda = x^{-2+\varepsilon}$ the variables ξ and ζ are both tending to zero as $x \to 0$. Thus we may use the well-known asymptotics of the Hankel functions near the origin while calculating Q through the formulae (8.26), (8.17), (8.23). As a result we have the expression for it

$$Q = \frac{1}{2\pi} \begin{pmatrix} A+B & , & Ae^{-\pi r/4} - Be^{\pi r/4} \\ Ae^{\pi r/4} - Be^{-\pi r/4} & , & A+B \end{pmatrix}, \tag{8.27}$$

where

$$A = 2^{3ir/2} e^{is/2} \Gamma^2\left(\frac{1}{2} + \frac{ir}{4}\right),$$

$$B = 2^{-3ir/2} e^{-is/2} \Gamma^2\left(\frac{1}{2} - \frac{ir}{4}\right).$$

Finally we recall the expression of Q through the monodromy parameters p, q and this would conclude the proof of the theorem.

We are now going to discuss the conditions on r and s providing the solution $u(x)$ being contained in the manifold $\mathbb{M}$. The comparison between the two formulae (8.27) and (1.58). expressing the matrix Q through r, s and p, q respectively, yields the constraint *)

$$1 + pq = \frac{4\pi^2}{(A+B)^2}. \tag{8.28}$$

It means that the solution (8.10) lies in the manifold $\mathbb{M}$ if the parameters r and s are such that

$$\operatorname{Im}(A+B) = 0. \tag{8.29}$$

---

*) Clearly the consztaint (8.28) may be obtained directly from the formulae (8.11).

It is worth noting that the condition (8.29) is obviously satisfied for all real-valued r and s (then $A = \bar{B}$) as well as for all pure imaginary r, s $(A, B \in \mathbb{R})$.

Thus under the constraint (8.29) we may able to describe exactly the asymptotics of solution $u(x)$ as $x \to +\infty$ with the initial condition (8.10). More precisely the theorem takes place

THEOREM 8.2. Suppose that the following statements are true for the solution $u(x)$:

a) the asymptotics (8.10) takes place near the origin while r and s satisfy the condition (8.29);

b) for all $x > 0$ the solution $u(x)$ has no singularities and its asymptotics at infinity is described by one of the formulae (8.4) - (8.7) with certain parameters α, β, φ.

Then the following formulae hold, connecting the parameter r, s of initial condition with that of α, β, φ

$$\alpha^2 = \mp \frac{16}{\pi} \ln \frac{1}{2\pi} \left| 2^{3ir/2} e^{is/2} \Gamma^2\left(\frac{1}{2} + \frac{ir}{4}\right) + 2^{-3ir/2} e^{-is/2} \Gamma^2\left(\frac{1}{2} - \frac{ir}{4}\right) \right|,$$

$$\beta^2 = \left| \frac{e^{\pi r/4} 2^{3ir/2} e^{is/2} \Gamma^2\left(\frac{1}{2} + \frac{ir}{4}\right) - e^{-\pi r/4} 2^{-3ir/2} e^{-is/2} \Gamma^2\left(\frac{1}{2} - \frac{ir}{4}\right)}{e^{is/2 - \pi r/4} 2^{3ir/2} \Gamma^2\left(\frac{1}{2} + \frac{ir}{4}\right) - e^{-is/2 + \pi r/4} 2^{-3ir/2} \Gamma^2\left(\frac{1}{2} - \frac{ir}{4}\right)} \right|, \qquad (8.30)$$

$$\varphi = \mp \frac{\alpha^2}{8} \ln 2 + \frac{3\pi}{4} - \arg \Gamma\left(\mp \frac{i\alpha^2}{16}\right) - \pi^{\frac{1-\varepsilon}{2}} -$$

$$- \arg\left\{ e^{is/2 - \pi r/4} 2^{3ir/2} \Gamma^2\left(\frac{1}{2} + \frac{ir}{4}\right) - e^{\pi r/4 - is/2} - 2^{-3ir/2} \Gamma^2\left(\frac{1}{2} - \frac{ir}{4}\right) \right\},$$

where the upper sign in „ $\mp$ " is taken for the case of asymptotics (8.4) and the lower one - for that of (8.5). The value of ε is

$$\varepsilon = \mathrm{sign}\left\{ 2^{3i\tau/2} e^{is/2} \Gamma^2\left(\frac{1}{2} + \frac{i\tau}{4}\right) + 2^{-3i\tau/2} e^{-is/2} \Gamma^2\left(\frac{1}{2} - \frac{i\tau}{4}\right) \right\} .$$

We shall espesially select the "separatrice" case of asymptotics (8.6), (8.7). The connection formulae here for τ, s and d, φ become extremely simple. For example, the solution (8.10) falls into the manifold $\mathbb{M}_p$ under the condition

$$\exp(is) = 2^{-3i\tau} \Gamma^2\left(\frac{1}{2} - \frac{i\tau}{4}\right) \Gamma^{-2}\left(\frac{1}{2} + \frac{i\tau}{4}\right) \exp\left(-\frac{\pi\tau}{2}\right) . \qquad (8.31)$$

Thus the formulae (8.11) become now

$$p = 2 \operatorname{sh} \frac{\pi\tau}{4} , \quad q = 0 .$$

This yields the connection formulae for the solution (8.6)

$$d = 2 \left| \operatorname{sh} \frac{\pi\tau}{4} \right| , \quad \varphi = \arg \operatorname{sh} \frac{\pi\tau}{4} , \qquad (8.32)$$

where τ, s are restricted by the constraints (8.29), (8.31).

Similarly for the solution (8.7) we have

$$d = 2 \left| \operatorname{sh} \frac{\pi\tau}{4} \right| , \quad \varphi = -\pi + \arg \operatorname{sh} \frac{\pi\tau}{4} , \qquad (8.33)$$

where alongside with (8.29) the additional condition holds

$$\exp(is) = 2^{-3i\tau} \Gamma^2\left(\frac{1}{2} - \frac{i\tau}{4}\right) \Gamma^{-2}\left(\frac{1}{2} + \frac{i\tau}{4}\right) \exp \frac{\pi\tau}{2} . \qquad (8.34)$$

In the particular case of $\mathrm{Re}\,\tau = 0$ the connection formulae (8.33) for the solution (8.7) were obtained earlier in [5] . Note by the way that for pure imaginary τ the conditions (8.34) and (8.29) are transformed into the explicit expressions of s through r :

$$\mathrm{Re}\, s = -\frac{i\pi r}{2} + 2\pi l, \quad l \in \mathbb{Z} \quad ,$$

$$\mathrm{Im}\, s = 3ir - 2\ln\left|\frac{\Gamma(\frac{1}{2} - \frac{ir}{4})}{\Gamma(\frac{1}{2} + \frac{ir}{4})}\right| \quad .$$

To our regret we are not able up till now to convert the assumption b) of the theorem 8.2 into its conclusion. The reason of this incompleteness lies in the lack of existence theorems for Painlevé equations in the framework of the inverse problem of monodromy theory. However the uniqueness theorem in the inverse problem does exist - it is reduced to the uniqueness of solution of the corresponding Riemann-Hilbert problem, which in its turn is the trivial corollary of the Liouville theorem. Therefore if we have some independent proof garanteing us the global solvability of the Cauchy problem for PIII equation, we may able to strenthen the statement of theorem 8.2 in any desirable way. For example such a proof does exist for the case of real-valued solutions. As a matter of fact any real-valued solution of PIII equation is obviously nonsingular as $x > 0$. Furthermore, as we have mentioned above the condition (8.29) is satisfied identically for real-valued r, s . On the other hand the real-valued solutions with the asymptotics (8.8) exist for any α and φ and the connection formulae (8.4) - (8.7) are trivially inverted giving the expressions of p, q through α, φ . Therefore we may formulate the stronger variant of theorem 8.2.

THEOREM 8.3. The real-valued solution of PIII equation (8.1) with the asymptotics (8.10) at the origin has the following asymptotics as $x \to +\infty$

$$u(x) = 2\pi k + \alpha x^{-1/2} \cos\{x - \frac{\alpha^2}{16}\ln x + \varphi\} + o(1) \quad , \tag{8.35}$$

where $\alpha > 0$,

$$\alpha^2 = -\frac{16}{\pi} \ln \frac{1}{\pi} \left| Re\{ 2^{3i\tau/2} e^{is/2} \Gamma^2(\tfrac{1}{2} + \tfrac{i\tau}{4}) \} \right| ,$$

$$\varphi = -\frac{\alpha^2}{8} \ln 2 + \frac{3\pi}{4} - \arg \Gamma(-\frac{i\alpha^2}{16}) -$$

$$-\frac{\pi}{2}\left(1 - \operatorname{sign}\{Re\left[2^{3i\tau/2} e^{is/2} \Gamma^2(\tfrac{1}{2} + \tfrac{i\tau}{4})\right]\}\right) -$$

$$-\arg\{e^{is/2 - \pi\tau/4} 2^{3i\tau/2} \Gamma^2(\tfrac{1}{2} + \tfrac{i\tau}{4}) - e^{-is/2 + \pi\tau/4} 2^{-3i\tau/2} \Gamma^2(\tfrac{1}{2} - \tfrac{i\tau}{4})\} \quad (8.36)$$

In order to complete our investigation of the Cauchy problem for PIII equation it remains to calculate the integer k in the formulae (8.4) - (8.7) through the parameters τ, s .

THEOREM 8.4. Under the assumptions of the theorem 8.2 the integer k in asymptotic formulae (8.4) - (8.7) is given by the following expression

$$k = \left[\frac{1}{2\pi} \{ \pi - s_1 + 3\tau_1 \ln 2 - 4 \arg \Gamma(\tfrac{1}{2} + \tfrac{i\tau}{4}) \} \right] , \quad (8.37)$$

where $s = s_1 + i s_2$, $\tau = \tau_1 + i\tau_2$ and the square brackets denote the entire part of a number.

PROOF. Consider the phase space (u, u_x) of the solutions to PIII equation. The points $u = 2\pi k$, $u_x = 0$ are stable focuses whereas $u = \pi + 2\pi k$, $u_x = 0$ are nonstable centers, being crossed by the separatrices, dividing the focuses. The real-valued case is presented at the figure 8.1

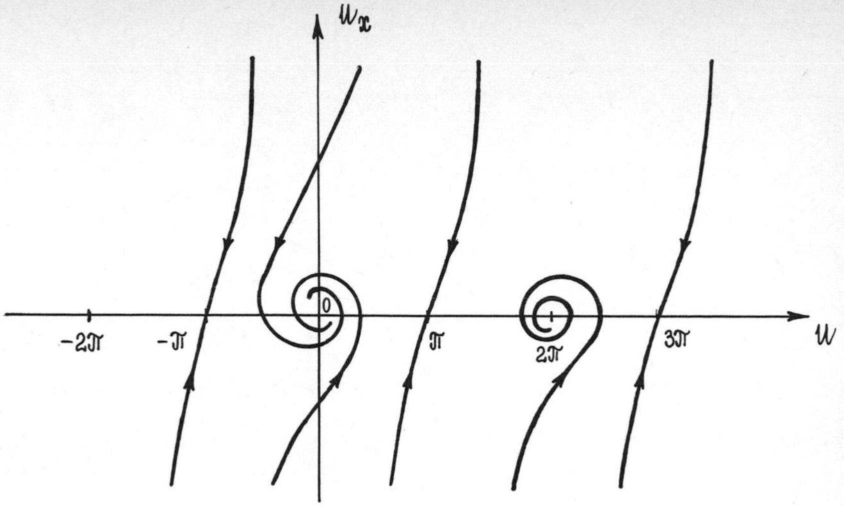

Figure 8.1

The asymptotic solutions (8.4) - (8.7) describe the behaviour of phase curve in the neighbourhood of the k-th focus, whereas the asymptotics (8.10)

$$u(x) = r \ln x + s + O(x^{2-|\mathrm{Im}\, r|})$$

$$u_x(x) = \frac{r}{x} + O(x^{1-|\mathrm{Im}\, r|}), \quad x \to 0,$$

corresponds to the solution at the infinity in (u, u_x) variables. The domains, divided by the separatrices, are numerated by different values of k, because there is only one focus $u = 2\pi k$, belonging to the k-th domain.

Let us derive equations for the separatrices expressing them through the initial data r, s. The integral curves described by asymptotics (8.4) - (8.7) fill in the whole domain in such a way that curves passing nearer to the boundary have the larger value of amplitude α in (8.4) - (8.7). Therefore the boundaries which coinside by definition with the separatrices correspond to the case of $\alpha^2 = \infty$. Then according to the theorem 8.2 we obtain from (8.30) (we remember that $|\operatorname{Im} \tau| < 2!$)

$$\alpha^2 = \infty \iff |A + B| = 0. \iff$$

$$\iff 2^{3i\tau}\,\Gamma^2\left(\tfrac{1}{2} + \tfrac{i\tau}{4}\right)\Gamma^{-2}\left(\tfrac{1}{2} - \tfrac{i\tau}{4}\right) = -e^{-is} . \tag{8.39}$$

This is just the complex-valued equation we need for the separatrices. We resolve it with respect to its modulus and argument and thus have two equations

$$s_2 = -3\tau_2 \ln 2 , \tag{8.40}$$

$$s_1 = \pi - 2\pi k + 3\tau_1 \ln 2 - 4 \arg \Gamma\left(\tfrac{1}{2} + \tfrac{i\tau}{4}\right) .$$

The value of integer k entering the second equation (8.40) has to be matched with the condition

$$k = 0 \iff \tau = s = 0 , \tag{8.41}$$

which follows just from the fact that $u = 0$ since $\tau = s = 0$ in (8.10). Bringing together (8.40) and (8.41) we arrive to the formula (8.37). The theorem is proved.

REMARK 8.4. It is clear from eq. (8.39) that the real-valued dimension of the separatrix manifold equals 2, whereas the dimension of the entire phase-space (u, u_x) is four. The apparent descrepansy in dimensions is explained by noting that in fact our solutions $u(x)$ belong to the manifold $\mathbb{M}$ which is three-dimensional due to the condition (8.29). Thus the codimension of separatrix equals to 1 on the manifold $\mathbb{M}$.

Chapter 9. THE MANIFOLD OF SOLUTIONS TO PAINLEVÉ II EQUATION INCREASING AS $x \to +\infty$. THE EXPRESSION OF THEIR ASYMPTOTICS THROUGH THE MONODROMY DATA. THE CONNECTION FORMULAE FOR PURE IMAGINARY SOLUTIONS

The qualitative analisis (see [34]) of PII equation shows that besides exponentially decreasing as $x \to +\infty$ solutions which have been investigated in Chapter 4, there exists a solution with the following asymptotic behaviour

$$u(x) = i\sqrt{\frac{x}{2}} + O(x^{-1/4}), \quad x \to +\infty . \tag{9.1}$$

In this chapter we are going to find out explicit parametrization of this solution in terms of the monodromy data p, q . For that purpose we exploit direct problem of the monodromy theory for the system (1.9) and construct its asymptotic solution under the assumption

$$\begin{aligned} u_0(x) &= O(x^{-1/4}) \ , \\ w_0(x) &= O(x^{+1/4}) \ , \end{aligned} \tag{9.2}$$

where

$$u = i\sqrt{\frac{x}{2}} + u_0 \ , \quad w = \frac{i}{2}\sqrt{\frac{1}{2x}} + w_0 = O(x^{1/4}). \tag{9.3}$$

Substitute (9.3) into the system (1.9) and rescale the variable λ :

$$z = x^{-1/2}\lambda .$$

The system on Ψ-function thus takes the form

$$\frac{d\Psi}{dz} = \tau A_0(z,\tau) \tag{9.4}$$

where

$$\tau = x^{3/2} \ ,$$

$$A_0 = -(4iz^2 - c)\sigma_3 - z(i2\sqrt{2} - 4id)\sigma_2 - \gamma\sigma_1 \ ,$$

$$c = 2\sqrt{\frac{2}{x}}\,u_0 - 2iu_0^2/x = O(\tau^{-1/2}) \ , \tag{9.5}$$

$$d = iu_0 x^{-1/2} = O(\tau^{-1/2}),$$

$$\gamma = 2w_0 x^{-1} + \frac{i}{\sqrt{2}}\,x^{-3/2} = O(\tau^{-1/2}) \ .$$

Now we are ready to apply WKB-technique to the system (9.4) in order to calculate the asymptotics of its monodromy data as $\tau \to \infty$. The procedure follows the scheme used in chapter 5 so we omit some details of calculations which coinside word by word with those made above.

The WKB ansatz for the solution of eq. (9.4) has again the structure (5.4) (note that $\tau \equiv x^{3/2}!$), where matrices T and Λ are defined as follows

$$\Lambda = -\mu\sigma_3 \ ,$$

$$\mu = 4i\sqrt{z^4 + \frac{1}{2}z^2 - \frac{c^2 + \gamma^2}{16}} \ , \tag{9.6}$$

$$T = \frac{1}{8i\sqrt{z^4 + \frac{1}{2}z^2}}\left[(4iz^2 - c + \mu)\sigma_0 + (4zd - 2\sqrt{2}z)\sigma_1 - i\gamma\sigma_2\right].$$

Assume that the branches of functions $\sqrt{z^4+\frac{1}{2}z^2}$ and $\mu(z)$ are fixed by the conditions

$$\sqrt{z^4+\frac{1}{2}z^2}\geqslant 0\ , \quad z\in\mathbb{R}\ ,$$

$$\mu(z)=4i\sqrt{z^4+\frac{1}{2}z^2}+o(1)\ ,\ \tau\to\infty\ .$$

The following three statements are obtained directly from (9.6) and the estimates (9.2):

a) There are two simple turning points

$$z_{\pm}=\pm\frac{i}{\sqrt{2}}$$

and one double turning point

$$z_0=0\ .$$

b) In the domain where

$$\sqrt{\tau}\ |z\sqrt{z^2+\tfrac{1}{2}}\,|\to\infty$$

the estimate holds for matrix T

$$T=\left(\frac{1}{2}+\frac{1}{2}z^2\left(z^4+\frac{1}{2}\right)^{-1/2}\right)\sigma_0-\frac{\sqrt{2}}{4i}\,z\left(z^4+\frac{1}{2}z^2\right)^{-1/2}\sigma_1+$$

$$+O\left(\frac{z+1}{\sqrt{\tau}\ \sqrt{z^4+\frac{1}{2}z^2}}\right)\ .$$

In particular it provides that

$$T\to I\ ,\quad z\to\infty\ ,\tag{9.7}$$

and

$$T=\left[\begin{pmatrix} 1 & 1 \\ i & -i \end{pmatrix}+o(1)\right]P_k \;,\quad z=O(\tau^{-1/2+\varepsilon})\;,\quad \varepsilon>0\;,$$

$$\arg z=\gamma_k\;,\quad k=1,4\;,$$

where

$$P_1=\begin{pmatrix} 1/2 & 0 \\ 0 & i/2 \end{pmatrix},\quad P_4=\begin{pmatrix} 0 & -i/2 \\ 1/2 & 0 \end{pmatrix},$$

$$\gamma_k=\begin{cases} 0\,, & k=1\,, \\ \pi\,, & k=4\,. \end{cases}$$

c) The conjugate Stokes lines tend asymptotically to the real axis, to the interval $\left[-\frac{i}{\sqrt{2}},\frac{i}{\sqrt{2}}\right]$ and to the branches of hyperbola

$$(\mathrm{Re}\,z)^2-\frac{1}{3}(\mathrm{Im}\,z)^2+\frac{1}{4}=0$$

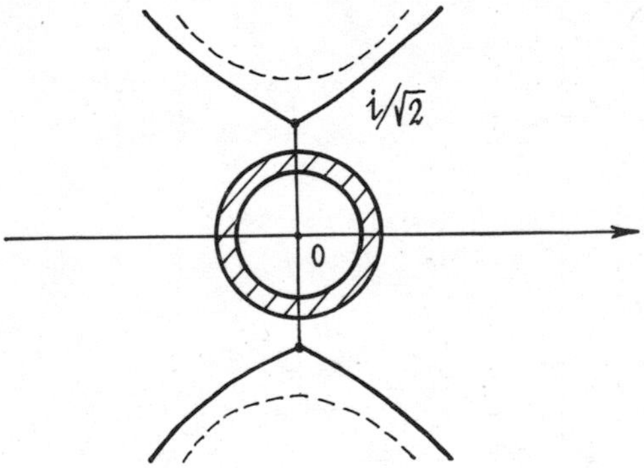

Figure 9.1. The solid lines are the limit positions of conjugate Stokes lines. Dotted lines are the branches of hyperbola $(\mathrm{Re}\,z)^2-\frac{1}{3}(\mathrm{Im}\,z)^2+\frac{1}{4}=0$. Shaded domain is the domain of matching Ω_0^ε.

The statements a) - c) permit us to outline two WKB-solutions

Ψ^1_{WKB} and Ψ^4_{WKB} with double asymptotics in τ and z :

$$\Psi^{K}_{ВКБ}(z) = \left[T(z) + o(1)\right] \exp\left\{-\tau\sigma_3 \int_0^z \mu(\eta)\, d\eta\right\} \tag{9.9}$$

$$\sqrt{\tau}\left|\sqrt{z^4 + \tfrac{1}{2}z^2}\right| \longrightarrow \infty\ , \quad \arg z = \gamma_K = \begin{cases} 0\ , & K = 1\ , \\ \pi\ , & K = 4\ . \end{cases}$$

It is clear from (9.7) and (9.9) that the solutions Ψ^1_{WKB} and Ψ^4_{WKB} are linked through the right diagonal matrices with the canonical solutions Ψ_1 and Ψ_4 . Moreover these matrices might be calculated explicitely through the formulae

$$C_K = \left[\Psi^{K}_{WKB}\right]^{-1} \Psi_K =$$

$$= \lim_{\substack{z \to \infty \\ \arg z = \gamma_K}} \left\{\tau\sigma_3 \int_0^z \mu(\eta)\, d\eta - \frac{4i}{3}\tau z^3 \sigma_3 - i\tau z \sigma_3\right\}. \tag{9.10}$$

At the neighbourhood of double turning point $z_0 = 0$ eq. (9.4) is simplified up to the system

$$\frac{d\Psi_0}{dz} = \tau \begin{pmatrix} c & -2\sqrt{2}\, z - \gamma \\ 2\sqrt{2}\, z - \gamma & -c \end{pmatrix} \Psi_0 \,. \tag{9.11}$$

If we put

$$\Psi_0 = \begin{pmatrix} 1 & 1 \\ i & -i \end{pmatrix} \Phi \quad ,$$

then the system (9.11) takes the well-acquainted form of (5.14),

where

$$v = e^{\frac{i\pi}{4}}\left(u_0(2x)^{1/4} + i w_0 (2x)^{-1/4}\right) = O(1) \quad ,$$

$$v^* = e^{\frac{i\pi}{4}}\left(u_0(2x)^{1/4} - i w_0 (2x)^{-1/4}\right) = O(1) \quad ,$$

$$\zeta = 2^{5/4} x^{3/4} z e^{-\frac{i\pi}{4}} \quad .$$

Therefore, under the usual assumption that $|v| \geqslant c > 0$ or $|v^*| \geqslant c > 0$, in the domain

$$\mathcal{D}_0^{\varepsilon} = \{ z : |z| \leqslant \tau^{-1/2+\varepsilon} \quad , \quad 0 < \varepsilon < \tfrac{1}{6} \}$$

there exists a solution $\Psi_0(z)$ to eq. (9.4) with the following asymptotic behaviour:

$$\Psi_0(z) = \begin{pmatrix} 1 & 1 \\ i\,, & -i \end{pmatrix} \begin{pmatrix} D_{-1-\nu}(i\zeta) \;, & D_\nu(\zeta) \\ \dot{D}_{-1-\nu}(i\zeta) \;, & \dot{D}_\nu(\zeta) \end{pmatrix} (I + o(1)) \quad , \qquad (9.12)$$

$$\tau \to \infty \;, \quad z \in \mathcal{D}_0^{\varepsilon} \;, \quad \arg z = 0, \pi \;,$$

where

$$\nu = -v v^* - 1 = -i\left(u_0^2 (2x)^{1/2} + w_0^2 (2x)^{-1/2}\right) - 1 \quad .$$

The asymptotic expansions of integral

$$\int_0^z \mu(\eta)\, d\eta$$

at infinity and at the circular domain

$$\Omega_0^{\varepsilon} = \{ z : |z| \approx \tau^{-1/2+\varepsilon} \quad , \quad 0 < \varepsilon < 1/6 \}$$

proceed just as in chapter 5, section III. As a result we obtain

$$\int_0^z \mu(\eta)d\eta = \frac{4i}{3}(z^2+\frac{1}{2})^{3/2} - i\sqrt{2}\,a^2 \ln\frac{\sqrt{z^2+\frac{1}{2}}-\frac{1}{\sqrt{2}}}{\sqrt{z^2+\frac{1}{2}}+\frac{1}{\sqrt{2}}} - i\frac{\sqrt{2}}{3} +$$

$$+ b + o(\tau^{-1}), \quad \tau\to\infty, \ |z|>C, \ \arg z=0, \tag{9.13}$$

and

$$\int_0^z \mu(\eta)\,d\eta = -\frac{1}{\tau}\Big(\frac{\zeta^2}{4} - (\nu+1)\ln\zeta + \frac{\nu+1}{2}\ln 8\sqrt{2}\,x^{3/2} -$$

$$-\frac{i\pi}{4}(\nu+1) + b + o(\tau^{-1}), \ \tau\to\infty, \ z\in\Omega_0^{\varepsilon}, \ \arg z=0, \tag{9.14}$$

where we have put

$$a^2 = \frac{u_0^2}{2x} + \frac{w_0^2}{4x^2} \quad ,$$

$$b = 4i\sqrt{2}\Big[\frac{a^2}{2}\ln(-a^2)^{1/2} - \frac{a^2}{4} - \frac{a^2}{2}\ln 2\Big] .$$

The estimates similar to (9.13), (9.14) in the case of $\arg z=\pi$ are derived using parity of the phase function $\mu(z)$

$$\mu(-z)=\mu(z), \quad z\in\mathbb{R} .$$

Hence putting $\arg z=\pi$ we get

$$\int_0^z \mu(\eta)d\eta = -\int_0^{-z}\mu(\eta)d\eta, \quad \arg(-z)=0 ,$$

which yields the asymptotic formulae

$$\int_0^z \mu(\eta)d\eta =$$

$$\int_0^z \mu(\eta)d\eta = -\frac{4i}{3}\left((-z)^2+\frac{1}{2}\right)^{3/2} + i\sqrt{2}\,a^2 \ln \frac{\sqrt{(-z)^2+\frac{1}{2}}-\frac{1}{\sqrt{2}}}{\sqrt{(-z)^2+\frac{1}{2}}+\frac{1}{\sqrt{2}}} + i\frac{\sqrt{2}}{3} -$$

$$-b + o(\tau^{-1}),\ \tau\to\infty,\ |z|>c,\ \arg z = \pi, \tag{9.15}$$

and

$$\int_0^z \mu(\eta)d\eta = \frac{1}{\tau}\Big[\frac{\zeta^2}{4} - (\nu+1)\ln\zeta + \frac{\nu+1}{2}\ln 8\sqrt{2}\,x^{3/2} +$$

$$+\frac{3\pi i}{4}(\nu+1) - b + o(\tau^{-1}),\ \tau\to\infty,\ z\in\Omega_0^\varepsilon,\ \arg z=\pi. \tag{9.16}$$

The substitution of asymptotics (9.13) and (9.15) into formula (9.10) leads to the equalities

$$C_1 = C_4^{-1} = \exp\left\{\left(-ix^{3/2}\frac{\sqrt{2}}{3} + x^{3/2}b\right)\sigma_3\right\} + o(1). \tag{9.17}$$

On the other hand the asymptotic expansions (9.14) and (9.16) together with (9.8), (9.9) and (9.12) produce the following asymptotic representations for the matrices $N_k = \Psi_0^{-1}\Psi_{WKB}^k$:

$$N_1 = \begin{pmatrix} \frac{1}{2}e^{\pi i(1+\nu)/2} & 0 \\ 0 & -\frac{i}{2}v \end{pmatrix} \exp\Big\{\sigma_3\Big(\frac{\nu+1}{2}\ln 8\sqrt{2}\,x^{3/2} -$$

$$-\frac{i\pi}{4}(\nu+1)\Big) - \tau b\sigma_3\Big\}(I+o(1)), \tag{9.18}$$

$$N_4^{-1} = \exp\left\{\sigma_3\left(\frac{\nu+1}{2}\ln 8\sqrt{2}\,x^{3/2} + \frac{3\pi i}{4}(\nu+1)\right) - \tau b\sigma_3\right\}(I+o(1))$$

$$\times \begin{pmatrix} -\frac{2i\sqrt{2\pi}}{v\Gamma(1+\nu)} e^{-\frac{\pi i}{2}(\nu+1)} & , & -\frac{2}{v} \\ 2ie^{\frac{3\pi i}{2}(\nu+1)} & , & 2i\sqrt{2\pi}\frac{e^{\pi i(\nu+1)}}{\Gamma(-\nu)} \end{pmatrix}. \qquad (9.19)$$

The monodromy data for the system (1.9) could be reconstructed through the matrix

$$S^{-1} = \begin{pmatrix} \frac{1+p^2}{1-pq} & \frac{p+q}{pq-1} \\ \frac{p+q}{pq-1} & \frac{1+q^2}{1-pq} \end{pmatrix} \equiv S_3^{-1} S_2^{-1} S_1^{-1} = \Psi_4^{-1}\Psi_1 = C_4^{-1} N_4^{-1} N_1 C_1 .$$

On the other hand the asymptotic expression of S^{-1} is given by the formulae (9.17) - (9.19) in the form

$$S^{-1} = \begin{pmatrix} -\frac{i}{v}\frac{\sqrt{2\pi}}{\Gamma(1+\nu)} e^{-\frac{2i\sqrt{2}}{3}x^{3/2} + (\nu+1)\ln 8\sqrt{2}\,x^{3/2} + \frac{\pi i}{2}(\nu+1)} & , ie^{i\pi(\nu+1)} \\ ie^{i\pi(\nu+1)} & , \frac{v\sqrt{2\pi}}{\Gamma(-\nu)} e^{\frac{2i\sqrt{2}}{3}x^{3/2} - (\nu+1)\ln 8\sqrt{2}\,x^{3/2} + \frac{i\pi}{2}(\nu+1)} \end{pmatrix} + o(1)$$

Hence the following theorem is true.

THEOREM 9.1. Let the parameters u and w in the system (1.9) depend on x such that

a)

$$u(x)=i\sqrt{\frac{x}{2}}+u_0\,,\qquad u_0=O(x^{-1/4})\,,$$

$$w(x)=\frac{i}{2}\sqrt{\frac{1}{2x}}+w_0\,,\qquad w_0=O(x^{1/4})\,. \tag{9.20}$$

b) either $|v|\geq c>0$, or $|v^*|\geq c>0$ as $x\to+\infty$, where

$$v=e^{\frac{i\pi}{4}}\left[(2x)^{1/4}u_0+i(2x)^{-1/4}w_0\right],$$

$$v^*=e^{\frac{i\pi}{4}}\left[(2x)^{1/4}u_0-i(2x)^{-1/4}w_0\right].$$

c) $$-1\leq \operatorname{Re} vv^*\leq 0\,.$$

Then the monodromy data p, q of the system (1.9) admit the following asymptotic representation

$$\frac{p+q}{pq-1}=ie^{-\pi\delta}\,,$$

$$\frac{1+p^2}{1-pq}=-\frac{i}{v}\frac{\sqrt{2\pi}}{\Gamma(i\delta)}\exp\left\{-\frac{2i\sqrt{2}}{3}x^{3/2}+i\delta\ln 8\sqrt{2}x^{3/2}-\frac{\pi}{2}\delta\right\}+o(1)\,, \tag{9.21}$$

$$\frac{1+q^2}{1-pq}=\frac{1}{v^*}\frac{\sqrt{2\pi}}{\Gamma(-i\delta)}\exp\left\{\frac{2i\sqrt{2}}{3}x^{3/2}-i\delta\ln 8\sqrt{2}x^{3/2}-\frac{\pi}{2}\delta\right\}+o(1)\,,$$

$$\delta=ivv^*.$$

REMARK 9.1. The first formula (9.21) is just the corollary of two others.

It is clear from the first equation (9.21) that the condition of isomonodromity of p, q data matches the estimates (9.20) if

$$i\,\frac{p+q}{1-pq}>0\ . \tag{9.22}$$

In particular this inequality holds under the condition $p=-\bar{q}$ (pure imaginary reduction) and $\operatorname{Im} p<0$. Let us introduce the submanifold $\mathbb{M}_{\mathrm{I}}^{-}$ of manifold $\mathbb{M}_{\mathrm{I}}$ (see Chapter 7):

$$\mathbb{M}_{\mathrm{I}}\supset\mathbb{M}_{\mathrm{I}}^{-}=\{(p,q)\in\mathbb{C}^{2}:p=-\bar{q},\ \operatorname{Im}p<0\}\ .$$

The theorem 9.1 guarantees that pure imaginary solutions of PII equation with the asymptotics (9.20) are contained in the manifold $\mathbb{M}_{\mathrm{I}}^{-}$. The explicit description of the second nontrivial term of their asymptotics is provided by the formulae (9.21):

$$u(x)=i\sqrt{\frac{x}{2}}+i(2x)^{-1/4}\rho\cos\left\{\frac{2\sqrt{2}}{3}x^{3/2}-\frac{3}{2}\rho^{2}\ln x+\theta\right\}+$$

$$+\,o\left(x^{-1/4}\right),\quad x\rightarrow+\infty\ , \tag{9.23}$$

where

$$\rho>0\ ,\quad \rho^{2}=\frac{1}{\pi}\ln\frac{1+|p|^{2}}{2\,|\operatorname{Im}p|}\ ,$$

$$\theta=-\frac{\pi}{4}+\frac{7}{2}\rho^{2}\ln 2-\arg\Gamma(i\rho^{2})-\arg(1+p^{2})\ . \tag{9.24}$$

As it was mentioned in Chapter 1 the tramsformation

$$u\longmapsto\bar{u}$$

correspond to the transformation of monodromy data

$$(p,q)\longmapsto(\bar{q},\bar{p})\ .$$

Therefore, introducing the submanifold

$$\mathbb{M}_I \supset \mathbb{M}_I^+ = \{(p,q)\in\mathbb{C}^2 :\ p=-\bar{q}\,,\ \mathrm{Im}\,p>0\}\,,$$

we obtain the parametrization of pure imaginary solutions of PII equation having the asymptotics

$$u(x) = -i\sqrt{\frac{x}{2}} + O(x^{-1/4})\,.$$

It takes the form

$$u(x) = -i\sqrt{\frac{x}{2}} - i(2x)^{-1/4}\rho\cos\left\{\frac{2\sqrt{2}}{3}x^{3/2} - \frac{3}{2}\rho^2\ln x + \theta\right\} +$$

$$+ o(x^{-1/4})\,, \qquad x \longrightarrow +\infty\,, \tag{9.25}$$

where the connection formulae for ρ and θ coinside with that of (9.24).

Let us discuss now the possibility of inversion of formulae (9.24). Assuming $\rho>0$ and $0\leqslant\theta<2\pi$ being arbitrary constant, denote

$$p = \zeta + i\eta\,,$$
$$\varkappa = e^{\pi\rho^2}\,, \tag{9.26}$$
$$\gamma = -\frac{\pi}{4} + \frac{7}{2}\rho^2\ln 2 - \arg\Gamma(i\rho^2) - \theta\,.$$

In order to determine the parameter p through ρ and θ consider the system of algebraic equations

$$\begin{cases} 1+\zeta^2+\eta^2 = 2\eta\varkappa\,, \\ \dfrac{1+\zeta^2-\eta^2}{2\zeta\eta} \equiv \mathrm{ctg}(\arg(1+p^2)) = \mathrm{ctg}\,\gamma\,. \end{cases} \tag{9.27}$$

We have to prove that for any $\varkappa > 1$ and $0 \leq \gamma < 2\pi$ the set of solutions to the (9.27) is nonempty and it contains only one element satisfying the condition

$$\eta > 0 , \quad \arg(1 + p^2) = \gamma . \tag{9.28}$$

Geometrically speaking the first equation (9.27) describes a circle lying in the upper halfplane and containing the point $(0, 1)$ for any $\varkappa > 1$. The curve of the second equation (9.27) is hyperbola with one of its branches lying in the upper halfplane and passing through the point $(0, 1)$. Hence for any $\varkappa > 1$ and $0 \leq \gamma < 2\pi$ the set of solutions to the (9.27) consists of two points. It is obvious that the condition (9.28) is satisfied only for one of these points - at the other one the second equation becomes $\arg(1 + p^2) = \gamma + \pi$. Therefore we have proved that formulae (9.24) might be uniqudy inverted for any $\rho > 0$, $0 \leq \theta < 2\pi$. It means that every solution of PII equation belonging to the manifold $\mathbb{M}_I^+$ is described by the formulae (9.25), (9.24) as $x \to +\infty$. Similarly, every solution from the manifold $\mathbb{M}_I^-$ has the asymptotics (9.23), where the connection formulae coinside with that of (9.24).

Let us return now to the results of Chapter 7, where it was found an asymptotic description as $x \to -\infty$ of solutions belonging to the manifold

$$\mathbb{M}_I = \mathbb{M}_I^+ \cup \mathbb{M}_I^{\circ} \cup \mathbb{M}_I^- .$$

Here $\mathbb{M}_I^{\circ} = \{ (p, q) : p = -\bar{q} , \operatorname{Im} p = 0 \}$ is separatrice manifold, which has been described asymptotically as $x \to \pm\infty$ in Chapter 7.

Bringing together *) these results with those obtained in the present chapter it becomes possible to write out the connection formulae linking together the asymptotics of solutions from $\mathbb{M}_I$ as $x \longrightarrow \pm\infty$:

THEOREM 9.2. Let $u(x)$ be an arbitrary pure imaginary solution of PII equation

$$u_{xx} = xu + 2u^3 .$$

Then the following statements are true:

a) There is an asymptotics as $x \longrightarrow -\infty$

$$u(x) = i\sqrt{2}\,\alpha(-x)^{-1/4}\sin\left\{\frac{2}{3}(-x)^{3/2} + \frac{3}{2}\alpha^2\ln(-x) + \varphi\right\} + o(x^{-1/4}) , \tag{9.29}$$

where the values $\alpha > 0$, $0 \leq \varphi < 2\pi$ might be arbitrary.

b) If the parameters α and φ of the solution $u(x)$ are linked by the constraint

$$\varphi = 3\alpha^2\ln 2 - \frac{\pi}{4} - \arg\Gamma(i\alpha^2) + \varepsilon\pi , \quad \varepsilon = 0, 1 , \tag{9.30}$$

then the solution $u(x)$ decreases exponentially as $x \to +\infty$:

$$u(x) = \frac{ia}{2\sqrt{\pi}}\, x^{-1/4} e^{-\frac{2}{3}x^{3/2}} \left(1 + o(1)\right) , \tag{9.31}$$

---

*) We remind that all the solutions here are pure imaginary. This fact guarantees the existence and smoothness of solutions to PII equation for all $x \in \mathbb{R}$.

where

$$a^2 = e^{2\pi\alpha^2} - 1 , \qquad \operatorname{sign} a = \begin{cases} 1, & \varepsilon = 0 , \\ -1, & \varepsilon = 1 . \end{cases}$$

c) If the constraint (9.30) fails (it is a general case!) then the solution grows up as $x \to +\infty$ in a power-like manner:

$$u(x) = \pm i\sqrt{\frac{x}{2}} \pm i(2x)^{-1/4} \rho \cos\left\{\frac{2\sqrt{2}}{3} x^{3/2} - \frac{3}{2}\rho^2 \ln x + \theta\right\} + \\ + o(x^{-1/4}) . \tag{9.32}$$

The asymptotics (9.32) remains true for any values of $\rho > 0$, $0 \leqslant \theta < 2\pi$, which just as α, φ , determine uniquely the solution $u(x)$. The parameters ρ and θ alongside with the choice of a sign in (9.32) are explicitely constructed through that of α and φ :

$$\rho^2 = \frac{1}{\pi} \ln \frac{1 + |p|^2}{2|\operatorname{Im} p|} , \tag{9.33}$$

$$\theta = -\frac{\pi}{4} + \frac{7}{2}\rho^2 \ln 2 - \arg\Gamma(i\rho^2) - \arg(1 + p^2) ,$$

where

$$p = \sqrt{e^{2\pi\alpha^2} - 1}\, \exp\left\{3i\alpha^2 \ln 2 - \frac{\pi i}{4} - i\arg\Gamma(i\alpha^2) - i\varphi\right\} .$$

The upper sign in (9.32) takes place when $\operatorname{Im} p < 0$.

REMARK 9.2. It is worth mentioning the curious of "oscillation pumting" in a general case solution. According to the formulae (9.33) it means that $\rho \to \infty$ as $\alpha \to 0$.

Chapter 10. THE MOVABLE POLES OF REAL-VALUED SOLUTIONS TO PAINLEVÉ II EQUATION AND THE EIGENFUNCTIONS OF ANHARMONIC OSCILLATOR

In this chapter we consider again the real-valued solutions to PII equation, belonging to the manifold $\mathbb{M}_R = \{(p,q):\ p=\bar{q},\ |p|<1\}$ introduced in Chapter 7. Remind that the solutions from $\mathbb{M}_R$ were parametrized there by an asymptotics as $x \to -\infty$ and the connection formulae were established between the asymptotic parameters and the monodromy data p and q. Our aim here is to investigate an asymptotic behaviour of solutions from $\mathbb{M}_R$ as $x \to +\infty$ and also to express the asymptotic parameters through the monodromy data. Thus, we'll establish the connection formulae linking the asymptotics of PII solutions on both infinities in a way similar to Ablowitz-Segur formulae (see Chapter 7).

Note at once a significant difference between asymptotic formulae constructed below and those derived in Chapter 7. The fact is that almost all the solutions from $\mathbb{M}_R$ have singularities (simple poles) at finite positive values of x, and the infinity appears to be their limit point. The only exeption is one-parameter family of solutions defined by the conditions $p=\bar{q}$, $\mathrm{Re}\,p=0$. The connection formulae for this family were found in Chapter 7.

The appearence of singularities tending to infinity makes impossible the construction of asymptotics as $x \to +\infty$. Instead of them we present asymptotic formulae for the distribution of poles at large x together with expressions for the Laurent series coefficients approximating the solution near the pole with large number. The solution in that case is interpreted as a meromorfic function satisfying the equation

$$u_{xx} = xu + 2u^3$$

for all $x \in \mathbb{C} \setminus \{a_1, a_2, \dots, a_n, \dots\}$, where $x = a_n$ are the poles of function $u = u(x)$. As far as the solution $u(x)$ is uniquely defined by its monodromy data ρ and q the coordinates of poles $x = a_n$ and the Laurent series coefficients are expressed uniquely through ρ and q. Thus the monodromy data determine the extension of solution $u(x)$ while passing over the singularity $x = a_n$ along the real axis.

The main tool for realization of our program would be again an asymptotic solution of the direct problem of the monodromy theory. We construct the Ψ-function as a solution of the system (1.9) under the assumption that the coefficients u, u'_x of (1.9) have the asymptotics

$$
\begin{aligned}
u(x) &= \frac{1}{x-a} - \frac{a}{6}(x-a) - \frac{1}{4}(x-a)^2 + b(x-a)^3 + O(x-a)^4, \\
u_x(x) &= -\frac{1}{(x-a)^2} - \frac{a}{6} - \frac{1}{2}(x-a) + 3b(x-a)^2 + O(x-a)^3, \\
& \qquad x \to a .
\end{aligned}
\tag{10.1}
$$

It is easy to check that the ansatz (10.1) satisfies the equation PII for all values of a, b. Moreover, as it was proved in the classical work [1] by Paul Painlevé, there exists a solution of PII equation which has the Laurent series expansion (10.1) in the neighbourhood of the point $x = a$.

Another characteristic feature of the result we are going to obtain is that the expressions of a, b through the monodromy data ρ, q would be constructed in term of solutions to a certain model equation. It is similar to the results of Chapter 8, where instead of WKB-approximations for the Ψ-function we have reduced the system (1.26) to the Bessel equation. Here the model equation appears to be the anharmonic oscillator

$$\frac{d^2V}{d\lambda^2} + (16\lambda^4 + 8a\lambda^2 - \nu^2)V = 0 \;, \qquad (10.2)$$

where $\nu^2 = -\frac{2}{9}a^2 - 40\,b$. The solution of the equation (10.2) satisfying the boundary condition

$$V(\lambda, a, \nu) \to \frac{1}{2\lambda} e^{4i/3\,\lambda^3 + i\lambda a} \;, \quad \lambda \to +\infty \,, \quad \operatorname{Im}\lambda = 0 \;.$$

presents the leading term of Ψ-finction as $x \to a$. As a result the parameters a, ν are linked with the monodromy data by means of the following scattering problem on the real axis

$$V(\lambda e^{i\pi}, a, \nu) =$$

$$= -\frac{2i\,\mathrm{Re}\,p}{1-|p|^2} V(\lambda, a, \nu) + \frac{1+\bar{p}^2}{|p|^2-1} \bar{V}(\lambda, a, \nu), \quad \lambda \to \infty \,.$$

In general there is no explicit solution of the inverse scattering problem, i.e. it is impossible to obtain an expression of a, ν through p in a closed form. However the problem admits an effective asymptotic solution in the case of large a . The situation is quite similar to those of Chapter 5, where the same problem has been solved for a regular solution $u(x)$. Finally we get the following formulae for asymptotic distribution of poles $x = a_n$ of the solutions to PII equation

$$(2a_n)^{3/2} = 6\pi n - \frac{3\varkappa}{4} \ln 6\pi n - \frac{3\varkappa}{2} \ln 2 - 3\arg(1+p^2) + \qquad (10.3)$$

$$+ 3\arg\Gamma\left(\frac{1}{2} + \frac{i\varkappa}{4}\right) + 3\pi + o(1)$$

$$\nu_n^2 = \varkappa \cdot (2a_n)^{1/2} + O(\tfrac{1}{n}) , \tag{10.4}$$

where $n \longrightarrow \infty$, $n \in \mathbb{Z}$,

$$\varkappa = \frac{4}{\pi} \ln \frac{2 \operatorname{Re} p}{1-|p|^2} , \quad |p| < 1 , \quad \operatorname{Re} p > 0 . \tag{10.5}$$

The formulae (10.3) - (10.5) make it possible to investigate the structure of solutions $u(x)$ from the manifold $\mathbb{M}_R$ as $x \to +\infty$. In general $u(x)$ is a meromorphic function with the infinite set of poles $x = a_n$ distributed on the real axis according to (10.3). For a fixed n the value a_n tends to infinity as $\varkappa \to -\infty$. The limit case of $\varkappa = -\infty$ when there are no poles, takes place due to (10.5) on a submanifold of $\mathbb{M}_R$ defined by the conditions

$$\operatorname{Re} p = 0 \; ; \quad |p| < 1 . \tag{10.6}$$

On the other hand the theorem 4.2 from Chapter 4 provides the existence of smooth decreasing at infinity solution just under the condition (10.6):

$$u(x) \sim |p| \cdot A_i(x) , \quad x \to +\infty .$$

Bringing together (10.3) - (10.5) and asymptotic formulae (7.9), parametrizing $\mathbb{M}_R$ as $x \to -\infty$, it is possible to obtain the connection formulae linking the asymptotic parameters d, φ as $x \to -\infty$ with that of a_n, ν_n as $x \to +\infty$.

I. ASYMPTOTIC EXPANSIONS FOR THE Ψ-FUNCTION.

We construct here an asymptotic solution of the system (1.9)

$$\frac{d\Psi}{d\lambda} = \left[-(4i\lambda^2 + ix + 2iu^2)\sigma_3 - 4u\lambda\sigma_2 - 2u_x\sigma_1 \right] \Psi \tag{1.9}$$

under the assumption, that coefficients u, u_x have the asymptotics

$$u(x)=\varepsilon^{-1}-\frac{a}{6}\varepsilon-\frac{1}{4}\varepsilon^{2}+b\varepsilon^{3}+O(\varepsilon^{4}), \tag{10.7}$$

$$u_x'(x)=-\varepsilon^{-2}-\frac{a}{6}-\frac{1}{2}\varepsilon+3b\varepsilon^{2}+O(\varepsilon^{3}),$$

$$\varepsilon=x-a\to 0 .$$

It is necessary as usual to find out only the leading term of Ψ-function as $\varepsilon\to 0$. Denote through $(v_1, v_2)^T$ the second column of Ψ and write out the equations on its components. Differentiating the system (1.9) in λ and retaining only the terms of order $O(\varepsilon^{-1})$, $O(1)$, $O(\varepsilon)$, we have

$$\frac{d^2 v_1}{d\lambda^2}+(16\lambda^4+8a\lambda^2-\nu^2)v_1 =$$

$$=-2\varepsilon\left[-\frac{dv_2}{d\lambda}+i(4\lambda^2+\frac{2a}{3})v_2+\frac{a}{3}v_1\right]+O(\varepsilon^2), \tag{10.8}$$

$$\frac{d^2 v_2}{d\lambda^2}+(16\lambda^4+8a\lambda^2-\nu^2)v_2 =$$

$$=2\varepsilon\left[\frac{dv_1}{d\lambda}+i(4\lambda^2+\frac{2a}{3})v_1-\frac{a}{3}v_2\right]+O(\varepsilon^2),$$

where $\nu^2=-\frac{2}{9}a^2-40b$.

Define the pair of functions $V, \overline{V}$ as a basis of solutions to the equation

$$\frac{d^2 V}{d\lambda^2}+(16\lambda^4+8a\lambda^2-\nu^2)V=0 \tag{10.9}$$

a scattering matrix S being defined through (10.20), we are able to reconstruct the Stokes matrices S_1, S_2, S_3 in a simple way. Taking into account a triangular structure of S_k and the constraint (1.21a) we immediatle get

$$S_2 = \begin{pmatrix} 1 & -(S)_{12} \\ 0 & 1 \end{pmatrix} .$$

The values p and q, entering S_1 and S_3 are calculated then by the formulae

$$q = -\frac{(S)_{22} - 1}{(S)_{12}} , \quad p = -\frac{(S_{11}) - 1}{(S)_{12}} .$$

We assume here, of course, that $(S)_{12} = r \neq 0$.

Consider now in detail the scattering problem for the anharmonic oscillator

$$\frac{d^2 V}{d\lambda^2} + (16\lambda^4 + 8a\lambda^2 - \nu^2)V = 0 . \tag{10.9}$$

Its basis of solutions we have defined above, putting the asymptotics as $\arg\lambda = 0$:

$$V(\lambda) \to \frac{1}{2\lambda} e^{4i\lambda^3/3 + ia\lambda} , \quad \lambda \to +\infty , \tag{10.10}$$

$$\bar{V}(\lambda) \to \frac{1}{2\lambda} e^{-4i\lambda^3/3 - ia\lambda} .$$

Hence on the other infinity we have

with the asymptotics

$$V(\lambda,a,\nu)\to\frac{1}{2\lambda}e^{4i/3\lambda^3+ia\lambda},$$
$$\overline{V}(\lambda,a,\nu)\to\frac{1}{2\lambda}e^{-4i/3\lambda^3-ia\lambda},\quad\lambda\to\infty,\ \arg\lambda=0\ . \tag{10.10}$$

The existence of solutions with given asymptotics is provided by the WKB-estimates of the form (see, for example, [38])

$$V\sim p^{-1/4}(\lambda)\exp i\int^{\lambda}\sqrt{p(\xi)}\,d\xi\ ,\quad\lambda\to\infty,$$

where $p(\lambda)=16\lambda^4+8a\lambda^2-\nu^2$.

The leading term of solutions v_1 , v_2 in (10.8) as $\varepsilon\to 0,\ \lambda\in\mathbb{R}$ we define as follows:

$$v_1=\frac{1}{\varepsilon}V(\lambda,a,\nu),$$
$$v_2=(2\lambda-\frac{1}{i\varepsilon})V(\lambda,a,\nu). \tag{10.11}$$

It is easy to check that the asymptotics (10.7), (10.10) yield the canonical boundary condition for the second column of Ψ matrix associated with the sector $\Omega_1=\{\lambda:-\frac{\pi}{3}<\arg\lambda<\frac{\pi}{3}\}$:

$$\begin{pmatrix}v_1\\ v_2\end{pmatrix}\to e^{4i/3\lambda^3+ia\lambda}\begin{pmatrix}\frac{u}{2\lambda}+o(\lambda^{-1})\\ 1+\frac{i}{2\lambda}(u_x^2-xu^2-u^4)+o(\lambda^{-1})\end{pmatrix}=$$

$$= e^{4i\lambda^3/3 + ia\lambda} \begin{pmatrix} \frac{1}{2\varepsilon\lambda} + o(\lambda^{-1}) \\ 1 + \frac{i}{2\varepsilon\lambda} + o(\lambda^{-1}) \end{pmatrix}, \quad \lambda \to \infty, \ \arg\lambda = 0 . \qquad (10.12)$$

The first column of Ψ_1-function may be obtained from (10.11) by the involution

$$v_1 \mapsto \bar{v}_2 \ , \quad v_2 \mapsto \bar{v}_1 \ .$$

Thus the canonical solution of the system (1.9) has the following asymptotics as $\varepsilon \to 0$, $\lambda \in \Omega_1$:

$$\Psi_1 = \frac{1}{\varepsilon} \begin{pmatrix} -i\bar{V}(\lambda), & V(\lambda) \\ \bar{V}(\lambda), & iV(\lambda) \end{pmatrix} + W_1(\lambda, \varepsilon), \ \arg\lambda = 0 , \qquad (10.13)$$

where the functions $V, \bar{V}$ are defined through (10.9), (10.10) and $W_1(\lambda, \varepsilon) = O(1)$.

Further we would be needed of another matrix Ψ_4 , defined in the sector

$$\Omega_4 = \left\{ \lambda : \frac{2\pi}{3} < \arg\lambda < \frac{4}{3}\pi \right\} .$$

Applying the involution (1.16)

$$\Psi(-\lambda) \mapsto \left[\Psi^T(\lambda) \right]^{-1}, \ \det\Psi = 1 ,$$

we get directly from (10.13)

$$\Psi_4 = \frac{1}{\varepsilon}\begin{pmatrix} iV(-\lambda) \; , \; -\bar{V}(-\lambda) \\ -V(-\lambda) \; , \; -i\bar{V}(-\lambda) \end{pmatrix} + W_4(\lambda,\varepsilon), \quad \arg\lambda = \pi \; , \qquad (10.14)$$

where $W_4(\lambda,\varepsilon) = O(1)$.

Establish now the range of applicability for the formal asymptotics (10.13), (10.14).

THEOREM 10.1. Let Ψ_1, Ψ_4 be the canonical solution to the system (1.9) in the sectors Ω_1, Ω_4. Here the boundary value condition (1.10) holds and the coefficients u, u_x have the asymptotics (10.7). Then the leading terms of Ψ_1, Ψ_4 have the form (10.13), (10.14) as $\varepsilon \to 0$ while the remainder terms

$$W_1 \, , \, W_4 = \exp\{i\sigma_3(\tfrac{4i}{3}\lambda^3 + ia\lambda)\}(1 + O(\lambda^{-1})) \; ,$$

$$\lambda \to \infty \; , \quad \operatorname{Im}\lambda = 0 \; . \qquad (10.15)$$

PROOF. The second column of Ψ_1 matrix may be presented due to (10.11), (10.13) in the form

$$v_1 = \frac{1}{\varepsilon} V + w_1 \; , \qquad (10.16)$$

$$v_2 = (2\lambda - \frac{1}{i\varepsilon})V + w_2 \; .$$

Then the remainder terms w_1, w_2 satisfy the system

$$\frac{d^2 w_1}{d\lambda^2} + p(\lambda) w_1 = 2\,[V' - 4i\lambda^2 V] + O(\varepsilon w_1) \; ,$$

$$\frac{d^2 w_2}{d\lambda^2} + p(\lambda) w_2 = -2\,[V' - 4i\lambda^2 V] + O(\varepsilon w_2) \; , \qquad (10.17)$$

which is obtained directly from (10.8) by substitution of the ansatz (10.16). Applying the asymptotics (10.10) one concludes that a function in square brackets of (10.17) vanishes as $\lambda \to \infty$

$$\frac{dV}{d\lambda} - 4i\lambda^2 V = O(\lambda^{-1} e^{4i\lambda^3/3 + ia\lambda}), \quad \arg\lambda = 0 .$$

The boundary value condition (10.12) together with ansatz (10.16) show that the functions w_1 , w_2 have to be vanishing as $\lambda \to \infty$. It is easy to prove the existence of such functions. In fact the first equation (10.17) with a condition

$$w_1 \longrightarrow 0 \ , \quad \lambda \longrightarrow \infty$$

is equivalent to the integral equation

$$w_1(\lambda) =$$

$$= \int_{\lambda}^{\infty} G(\lambda - \mu) \left[2(V' - 4i\mu^2 V) + O(\varepsilon w_1(\mu)) \right] d\mu ,$$

where $G(\lambda)$ is the Green function for equation (10.2)

$$\frac{d^2 G}{d\lambda^2} + p(\lambda) G = \delta(\lambda)$$

with the conditions $G(\lambda) = 0$ as $\lambda < 0$ and

$$G(\lambda) \to 0 \ , \quad \lambda \to \infty . \tag{10.18}$$

Estimations of integral in the right-hand side for $\lambda \to \infty$ together with a perturbation method for small ε lead to the asymptotics

$$|w_1|, |w_2| = O(\lambda^{-1}), \qquad \lambda \to \infty ,$$

which yield the estimates (10.15).

The first column of Ψ_1 and the matrix Ψ_4 are considered in a similar way. The theorem in proved.

REMARK 10.1. The asymptotic expressions (10.13), (10.14) for canonical solutions Ψ_1, Ψ_4 remain true only for real-valued λ since the exponents of $V, \bar{V}$ are bounded as $\lambda \in \mathbb{R}, \lambda \to \infty$. Expanding the functions Ψ_1, Ψ_4 into the sectors Ω_1, Ω_4 respectively one meets the well-known Stokes phenomenon, described in Chapter 1. As a result the remainder terms in (10.13), (10.14) become the leading ones at the rays $\arg\lambda = \frac{\pi}{3}$ and $\arg\lambda = \frac{4\pi}{3}$. They determine thus the Stokes multipliers through the canonical asymptotics (1.14) at infinity. Through the leading term functions $V, \bar{V}$ vanish at infinity, their Stokes multipliers would coinside with those defined for the canonical Ψ-functions because of the large parameter ε^{-1} entering the leading terms of (10.13), (10.14) (see theorem 10.2 below).

II. THE STOKES MATRICES FOR ANHARMONIC OSCILLATOR

We establish here a relation between the Stokes matrices for the equation (10.9) and that of the system (1.9). It would be shown that the former are just the limit case of the latter as $\varepsilon \to 0$ and they are constructed in terms of a certain scattering problem for (10.9) through a suitable choice of parameters a, ν.

Remind first that the canonical solutions Ψ_k of the system (1.9) are linked with each other by the constraints (see Chapter 1)

$$\Psi_{k+1} = \Psi_k S_k , \quad k=1,2,\dots,6 ,$$

where the Stokes matrices are defined by (1.19). This yields the formal equation on Ψ_4 and Ψ_1 (see the Chapter 9)

$$\Psi_1 = \Psi_4 S ,$$

where

$$S = S_3^{-1} S_2^{-1} S_1^{-1} = \frac{1}{1-pq} \begin{pmatrix} 1+p^2 & -p-q \\ -p-q & 1+q^2 \end{pmatrix} .$$

According to the conditions $p = \bar{q}$, $|p| < 1$ defining the manifold $\mathbb{M}_R$ the latter formula take the form

$$S = \frac{1}{1-|p|^2} \begin{pmatrix} 1+p^2 & -p-\bar{p} \\ -p-\bar{p} & 1+\bar{p}^2 \end{pmatrix} . \tag{10.19}$$

The formal identity $\Psi_4 = \Psi_1 S^{-1}$ one can interpret as a scattering problem on the real axis for the system (1.9)

$$\begin{aligned} \Psi \to \Psi_4 \Big|_{\lambda \to -\infty} &= \exp\{i\sigma_3 (\tfrac{4}{3}\lambda^3 + \lambda x)\} , \\ \Psi \to \Psi_1 \Big|_{\lambda \to +\infty} \cdot S^{-1} &= \exp\{i\sigma_3 (\tfrac{4}{3}\lambda^3 + \lambda x)\} S^{-1} . \end{aligned} \tag{10.20}$$

In fact the matrices Ψ_1 , Ψ_4 represent the solutions of (1.9), normalized according to (1.14) as $\lambda \to +\infty$ and $\lambda \to -\infty$. While passing along a certain contour λ (in the upper half-plane) from Ω_4 to Ω_1 the matrix Ψ_4 transforms into $\Psi_1 \cdot S^{-1}$ with a certain S being constant in λ . We have the equations $\Psi_{k+1} = \Psi_k S_k$ on the boundaries of sectors Ω_k , $k = 3, 2, 1$ so the matrix S turns to be equal to that of (10.19). Inversely, having

$$V(\lambda) \to \frac{1}{2\lambda}\left\{ Pe^{4i\lambda^3/3 + ia\lambda} + Qe^{-4i\lambda^3/3 - ia\lambda} \right\},$$

$$\bar{V}(\lambda) \to \frac{1}{2\lambda}\left\{ \bar{Q}e^{4i\lambda^3/3 + ia\lambda} + \bar{P}e^{-4i\lambda^3/3 - ia\lambda} \right\}, \lambda \to -\infty, \tag{10.21}$$

due to the self-adjointness of equation (10.9). The scattering data satisfies usual condition of unitarity

$$|P|^2 - |Q|^2 = 1 .$$

In terms of the scattering matrix

$$S_0 = \begin{pmatrix} \bar{P} & iQ \\ -i\bar{Q} & P \end{pmatrix} \tag{10.22}$$

the equation (10.21) may be written as follows

$$\left(-i\bar{V}(\lambda), V(\lambda)\right) = \left(iV(-\lambda), -\bar{V}(-\lambda)\right) S_0 . \tag{10.23}$$

The following theorem provides a comparison between S_0 (10.22) and S (10.19)

THEOREM 10.2. If the coefficients u, u_x of the system (1.9) satisfy an asymptotics (10.7), then for matrices S_0, S defined by (10.23), (10.19) respectively, the estimate holds

$$S - S_0 = O(\varepsilon) .$$

PROOF. According to the theorem 10.1 there is a solution Ψ satisfying an asymptotic (10.20) which may be represented as (10.13) (10.14) on the real axis. By substituting them into (10.20) one obtains

$$\left\{ \frac{1}{\varepsilon}\left(-i\bar{V}(\lambda), V(\lambda)\right) + O(1) \right\} =$$

$$=\left\{\frac{1}{\varepsilon}(iV(-\lambda), -\bar{V}(-\lambda)) + O(1)\right\} S .$$

Comparing the leading terms here (of order $O(\varepsilon^{-1})$) with the scattering equation (10.23) we prove their identity.

COROLLARY 1. The scattering data P , Q in (10.21) are related to the monodromy data p , $q = \bar{p}$ by the formulae

$$P = \frac{1+\bar{p}^2}{1-|p|^2} + O(\varepsilon) , \quad Q = i\,\frac{p+\bar{p}}{1-|p|^2} + O(\varepsilon). \tag{10.24}$$

COROLLARY 2. Let $(S)_{12} \neq 0$, then the Stokes matrices S_k for the system (1.9) with coefficient u, u_x of the form (10.7) satisfy the estimates

$$S_k - S_{k0} = O(\varepsilon) , \quad k=1,2,3 ,$$

where S_{k0} are corresponding Stokes matrices for the anharmonic oscillator (10.9).

PROOF is obtained directly from the estimate (10.24)and the formulae

$$S_1 = \begin{pmatrix} 1 & 0 \\ -\frac{(S)_{11}-1}{(S)_{12}} & 1 \end{pmatrix}, \quad S_2 = \begin{pmatrix} 1 & -(S)_{12} \\ 0 & 1 \end{pmatrix},$$

$$S_3 = \begin{pmatrix} 1 & 0 \\ -\frac{(S)_{22}-1}{(S)_{12}} & 1 \end{pmatrix}.$$

THEOREM 10.3. Let $u = u(x)$ be real-valued solution to the PII equation. It is fixed by the monodromy data $p, q=\bar{p}$, $|p| < 1$ and belongs to the manifold M_R . Then there exist the real-values a , $b = -\frac{1}{40}(\nu^2 + \frac{2}{9}a^2)$, defined through the scattering problem (10.23), (10.24), such that $u(x)$ has the Laurent expansion (10.7) with the parameters a, b .

PROOF. We construct first the solution of a scattering problem (10.23) for anharmonic oscillator (10.9) with the monodromy data (10.24)

$$P = \frac{1+\bar{\rho}^2}{1-|\rho|^2}, \quad Q = \frac{\rho+\bar{\rho}}{1-|\rho|^2} i .$$

The existence and uniqueness of solution to the inverse scattering problem for the equation (10.9) may be established by an usual transition to the Gelfand-Levitan-Marchenko integral equation. By solving it we obtain the values a, ν entering a potential and also the Jost function V .

We apply now the classical result [1] by Paul Painlevé about the local existence of a precise solution $\hat{u}(x)$ of PII equation, which has the Laurent series expansion (10.7) with the prescribed parameters a, ν . Substitute then the solution $\hat{u}(x)$ into the system (1.9) and calculate an asymptotics of Ψ-function as $\varepsilon \to 0$. We get thus the Stokes matrices $\hat{S}_k$ satisfying due to the Corollary 2 of theorem 10.2 the following estimates

$$\hat{S}_k - S_k = \hat{S}_k - S_{k0} = O(\varepsilon),$$

where S_k are the Stokes matrices associated with the initial solution $u(x)$. The isomonodromic condition takes place for the solution $\hat{u}(x)$, provided by the theorem 3.1 of Chapter 3. It means that all S_k are independent of x , i.e. of $\varepsilon = x - a$. This yields immediately $\hat{S}_k = S_k$ and thus $\hat{u}(x)$ coinsides with $u(x)$ in the neighbourhood of $x = a$. The theorem is proved.

III. THE ASYMPTOTIC DISTRIBUTION OF POLES FOR A SOLUTION OF PII EQUATION

The inverse scattering problem (10.3), i.e. the calculation of a, ν through the given p, q, seems to have no exact solution for any finite a, ν. As a matter of fact a point $x=a$ is a pole of the function

$$u(x) = -2i \lim_{\lambda\to\infty} \lambda V(\lambda, a, \nu) e^{-4i\lambda^3/3 - ia\lambda},$$

whereas V is determined via the Gelfand-Levitan-Marchenko equation with non-degenerate kernel. However, the asymptotics of V can be calculated explicitly for $a \to \infty$, and its parameters are expressed exactly through the monodromy data p and q. Thus an explicit formula for the distribution of poles would be obtained.

We look for the asymptotics of V as $a \to +\infty$ in the form of WKB-approximation

$$V \sim C \rho^{-1/4}(\lambda) \exp\left\{\pm i \int^{\lambda} \sqrt{\rho(\xi)}\, d\xi\right\}, \qquad (10.25)$$

where $\rho(\lambda) = 16\lambda^4 + 8a\lambda^2 - \nu^2$.

Note that WKB- solutions for anharmonic oscillator (10.9) were studied in detail in [54].

The asymptotics (10.25) appears to be double, i.e. the formula (10.25) represents the leading term of V both for $a\to\infty$ and $\lambda\to\infty$. Therefore the scattering problem (10.21) may be written down in the form

$$V(\lambda) \sim C \begin{cases} \rho^{-1/4}(\lambda) \exp\left\{i \int_{\mu_0}^{\lambda} \sqrt{\rho(\xi)}\, d\xi\right\}, & \lambda > |\mu_0| \\ \rho^{-1/4}(\lambda) \left[P_1 \exp\left\{i \int_{-\mu_0}^{\lambda} \sqrt{\rho(\xi)}\, d\xi\right\} + \right. & \end{cases}$$

$$+ Q_1 \exp\left\{-i\int_{-\mu_0}^{\lambda} \sqrt{p(\xi)}\,d\xi\right\}\Big], \quad \lambda < -|\mu_0| . \tag{10.26}$$

The coefficients C, P_1, Q_1 in (10.26) would be determined further. The points denoted via μ_0, $-\mu_0$ are supposed to be the turning points, i.e. the zeroes of potential $p(\lambda)$ lying at a small distance (of order $O(a^{-1/4})$) from each other. More precisely, we assume that

$$\nu^2 = \varkappa\sqrt{2a} + O(a^{-1}), \quad a \to \infty, \quad \varkappa \in \mathbb{R} \tag{10.27}$$

and put

$$16\lambda^4 + 8a\lambda^2 - \nu^2 = (4\lambda^2 - 4\mu_0^2)(4\lambda^2 + \sigma^2),$$

where

$$\mu = \frac{\sqrt{\varkappa}}{2(2a)^{1/4}}\left(1 + o(1)\right), \quad \sigma^2 = 2a + \frac{\varkappa}{\sqrt{2a}} + \cdots .$$

The branches of radicals $p^{1/2}(\lambda)$, $p^{-1/4}(\lambda)$ are chosen in such a way, that

$$p^{1/2}(\lambda), \; p^{-1/4}(\lambda) \geqslant 0, \quad \lambda \in \mathbb{R}, \; |\lambda| > |\mu_0| .$$

Then we have

$$p^{1/2}(-\lambda) = p^{1/2}(\lambda),$$

$$p^{-1/4}(-\lambda) = p^{-1/4}(\lambda), \quad \lambda \in \mathbb{R}, \; |\lambda| > |\mu_0| .$$

It is clear that the remoted turning points $\pm i\sigma \sim \sqrt{2a}$ provide an exponentially small inset to the scattering amplitudes along the real axis, so we do not take them into account in our analysis of the leading terms of asymptotics.

We need now an asymptotics of phase integrals in (10.26) as $\lambda \to \infty$ and $\lambda \to 0$. Due to the choice of the branches for $\sqrt{p(\xi)}$ it is sufficient to evaluate the integral

$$I = \int_{\mu_0}^{\lambda} \sqrt{p(\xi)}\, d\xi = \int_{\mu_0}^{\lambda} \sqrt{16\xi^4 + 8a\xi^2 - \nu^2}\, d\xi \ , \quad \lambda > |\mu_0| .$$

We note that

$$\int_{\mu_0}^{\lambda} \sqrt{16\xi^4 + 8a\xi^2 - \nu^2}\, d\xi = \int_{0}^{\lambda} \sqrt{16\xi^4 + 8a\xi^2 - \nu^2}\, d\xi - \frac{i\pi æ}{8} + o(1) ,$$

$$a \to \infty$$

and

$$\int_{0}^{\lambda} \sqrt{16\xi^4 + 8a\xi^2 - \nu^2}\, d\xi = \frac{a^{3/2}}{2} \int_{0}^{z} \sqrt{\eta^4 + 2\eta^2 + \alpha}\, d\eta$$

were

$$\eta = \frac{2\xi}{\sqrt{a}} \ , \quad z = \frac{2\lambda}{\sqrt{a}} \ , \quad \alpha = -\frac{\nu^2}{a^2} = -\frac{\sqrt{2}\, æ}{a^{3/2}} .$$

The needful asymptotics of the integral $\int_0^z \sqrt{\eta^4 + 2\eta^2 + \alpha}\, d\eta$ has been constructed in Chapter 9. Then we have following estimates for the integral I :

$$1.\quad I = \frac{4}{3}\lambda^3 + a\lambda - \frac{(2a)^{3/2}}{6} - \frac{\varkappa}{4}\ln\frac{4(2a)^{3/4}}{\sqrt{\varkappa}} - \frac{\varkappa}{8} + o(1) \,,$$

$$a \to \infty \,, \quad |\lambda| > c \,, \quad \arg\lambda = 0 \,. \tag{10.28}$$

$$2.\quad I = \frac{\zeta^2}{4} - \frac{\varkappa}{4}\ln\zeta + \frac{\varkappa}{4}\ln\frac{\sqrt{\varkappa}}{2} - \frac{\varkappa}{8} + o(1) \,, \tag{10.29}$$

$$\zeta = 2(2a)^{1/4}\lambda \,, \quad a \to \infty \,, \quad \lambda = O(a^{-1/4+\varepsilon}),\, 0<\varepsilon<1/6,\, \arg\lambda = 0.$$

By substituting (10.28) into (10.26) and comparing with (10.21) we conclude that

$$C = e^{-iA} \,, \quad Q_1 = -Q \,, \quad P_1 = -Pe^{2iA} \,,$$

where

$$A = -\frac{\varkappa}{4}\ln\frac{4(2a)^{3/4}}{\sqrt{\varkappa}} - \frac{(2a)^{3/2}}{6} - \frac{\varkappa}{8} \,,$$

P, Q are the scattering data in (10.21).

In a neighbourhood $(|\lambda| \leq a^{-1/4+\varepsilon})$ of the turning points the equation (10.9) takes the form

$$\frac{d^2V}{d\zeta^2} + \frac{1}{4}(\zeta^2 - \varkappa)V = 0 \,. \tag{10.30}$$

Its solutions are the parabolic cylinder functions

$$V(\zeta) = C_1 D_{-1/2 - i\varkappa/4}(\zeta e^{-i\pi/4}) + C_2 D_{-1/2 - i\varkappa/4}(-\zeta e^{-i\pi/4}) \,.$$

According to the first equation (10.26) together with asymptotics (10.29) we have a matching condition as $\zeta \to +\infty \, (0 < \lambda \sim a^{-1/4+\varepsilon})$

$$V(\zeta) = \mathcal{D} \cdot \zeta^{-1/2 - i\varkappa/4} \cdot e^{i\zeta^2/4} \left[1 + O(\zeta^{-1}) \right] \tag{10.31}$$

where

$$\mathcal{D} = (2a)^{-1/8} \cdot e^{-iA - iB} ,$$

$$B = \frac{\varkappa}{8} - \frac{\varkappa}{4} \ln \frac{\sqrt{\varkappa}}{2} .$$

Comparing (10.31) with the well-known asymptotics of the parabolic cylinder functions (see, for example, [53]), we conclude that a solution of (10.30) have to be taken in the form

$$V(\zeta) = C_1 D_{-1/2 - i\varkappa/4} (\zeta e^{-i\pi/4}) , \tag{10.32}$$

where the matching constant C_1 appears to be

$$C_1 = \mathcal{D} e^{\pi\varkappa/16 - \pi i/8} = (2a)^{-1/8} e^{\pi\varkappa/16 - \pi i/8 - iA - iB} . \tag{10.33}$$

Extending the solution (10.32) from right to left along the real axis we write out its asymptotics as $\zeta \to -\infty$ and match it with that of (10.26) as $0 > \lambda \sim a^{1/4+\varepsilon}$. As a result there would be equations on the coefficients before the leading terms $\zeta^{-1/2 - i\varkappa/4} e^{i\zeta^2/4}$ $\zeta^{-1/2 + i\varkappa/4} e^{-i\zeta^2/4}$:

$$Q \cdot (2a)^{-1/8} e^{-iA - iB} = i C_1 e^{3\pi\varkappa/16 + \pi i/8} ,$$

$$p\cdot(2a)^{-1/8}e^{iA+iB}=-C_1\frac{\sqrt{2\pi}\,e^{\pi\varkappa/16+\pi i/8}}{\Gamma\left(\frac{1}{2}+\frac{i\varkappa}{4}\right)} \quad . \tag{10.34}$$

The first equation (10.34) yields

$$e^{\pi\varkappa/4}=-iQ=\frac{p+\bar{p}}{1-|p|^2} \quad ,$$

so that

$$\varkappa=\frac{4}{\pi}\ln\frac{p+\bar{p}}{1-|p|^2} \quad , \quad \operatorname{Re}p>0 \quad . \tag{10.35}$$

The unitarity condition $|p|^2-|Q|^2=1$ provides the coincidence of absolute values of the right and left sides in the second equation (10.34). Equating thus the arguments in (10.34) we have

$$\frac{(2a)^{3/2}}{3}+\frac{\varkappa}{2}\ln 2(2a)^{3/4}-\arg\Gamma\left(\frac{1}{2}+\frac{i\varkappa}{4}\right)+\arg(1+p^2)=2\pi n+\pi \; ,$$

where n is an integer, $n\to\infty$ since we have assumed $a\to\infty$. Solving the latter equation for a, we obtain the asymptotic formula (10.4). Thus we have proved the following

THEOREM 10.4. Let $u\in \mathbb{M}_R$ be the real-valued solution of PII equation, fixed by the monodromy data p, q :

$$p=\bar{q} \; , \; |p|<1 \; , \; \operatorname{Re}p>0 \; .$$

Then it has an infinite number of real poles with the asymptotic distribution (10.4) as $x \to +\infty$. In the special case of $\mathrm{Re}\,p = 0$ the poles are absent and the solution $u(x)$ turns to be smooth exponentially decreasing as $x \to +\infty$ with the asymptotics described in the theorem 4.2.

REMARK 10.2. As it easy to understand,

$$u \to -u \Leftrightarrow p \to -p .$$

Then the solution to PII equation for which

$$p = \bar{q} \ , \ |p| < 1 \ , \ \mathrm{Re}\,p < 0$$

possess the following behaviour as $x \to +\infty$:

$$u(x) = -\frac{1}{x-a_n} + \frac{a_n}{6}(x-a_n) + \frac{1}{4}(x-a_n)^2 - b_n(x-a_n)^3 + \cdots$$

where

$$(2a_n)^{3/2} = 6\pi n - \frac{3\varkappa}{4}\ln 6\pi n - \frac{3\varkappa}{2}\ln 2 - 3\arg(1+p^2) +$$

$$+ 3\arg\Gamma\left(\frac{1}{2} + \frac{i\varkappa}{4}\right) + 3\pi + O\left(\frac{\ln n}{n}\right), \ n \to \infty$$

$$\varkappa = \frac{4}{\pi}\ln - \frac{2\,\mathrm{Re}\,p}{1-|p|^2} \ .$$

REMARK 10.3. We have essentially used in this Chapter as a main technical tool a reduction of the monodromy theory direct problem for the system (1.9) to a certain scattering problem for the anharmonic oscillator (10.9). However, an asymptotics of the Ψ-function as well as the Stokes matrices might be calculated as $\varepsilon \to 0$ in a quite different manner, than we have done here by the ansatz (10.13), (10.14). Those calculations, being carried out independently by A.A. Kapaev (see Appendix 2) produce the same result as we have obtained here for a distribution of poles $x = a_n$ as $x \to \infty$. Moreover the method proposed by A.A.Kapaev works also in that case when the scattering matrix S for anharmonic oscillator becomes trivial, whereas the Stokes matrices S_k are non-trivial and contain pure imaginary parameter $p, |p| > 1$. The case corresponds to a smooth exponentially decreasing solution $u(x)$ with the asymptotics

$$u(x) \sim ip \cdot Ai(x), \quad x \to +\infty, \quad |p| > 1.$$

This solutions becomes singular as $x \to -\infty$ with a set of poles of the form (10.7), where $a < 0$. The distribution of the poles as $a \to -\infty$ found out by A.A.Kapaev, is given in the Appendix 2. The proof of the asymptotic formulae we are going to omit in the main text, trying not to overcomplicate it.

Chapter 11. THE MOVABLE POLES OF THE SOLUTIONS OF PAINLEVÉ III EQUATION AND THEIR CONNECTION WITH MATHIFŪ FUNCTIONS

We shall consider here another special kind of PainlevéIII equation:

$$u_{xx} + \frac{1}{x} u_x = 4 \sinh u , \qquad x > 0 . \tag{11.1}$$

It is transformed to eq. (3.5) by simple change of variables $u \mapsto \pi + iu$. The equation (11.1) arises in many context of modern mathematical physics (see, for example, [5], [9], [11]). Some applications suggest to study not only smooth bounded solutions of eq. (11.1), but also the singularities, which appear to be poles of the solutions of Painlevé equation been written down in a canonical way (see below eq. (11.4)).

We suppose the solution of eq. (11.1) $u = u(x)$ to be real-valued function with the following asymptotics near the origin

$$u(x) = r \ln x + s + O(x^{2-|r|}), \; x \to 0, \; |r| < 2. \tag{11.2}$$

It is shown in [5] that under a special choice of initial data r, s there exists a smooth solution of eq. (11.1) with asymptotics

$$u(x) \sim \alpha x^{-1/2} e^{-2x} , \qquad x \to \infty . \tag{11.3}$$

The exact connection formulas are given in [5] for the parameters r, s and α (see below eq. (11.11), (11.12)). Taking linear limit as $r, s \to 0$ in the solution (11.3) we get Bessel function of imaginary argument $K_0(2x)$. The nonlinearity of eq. (11.1) leaves no room for the corresponding $I_0(2x)$ function, i.e. the solution with an exponential growth-rate at infinity. It is easy to prove that any solution, linearly independent to that of (11.3) must have singularities at finite values of x. These singularities correspond to the movable poles of the solution of Painlevé III equation, presented in the canonical form

$$w_{xx} = \frac{w_x^2}{w} - \frac{1}{x} w_x + w^3 - \frac{1}{w} . \tag{11.4}$$

It is obtained directly from eq. (11.1) under the transform

$$u = 2 \ln w . \tag{11.5}$$

The word "movable" used above means that the coordinates of poles depend only of initial data.

The logarithmic transform (11.5) translates the poles of w into the branching points of u. So near the singularity $x=a$ we would seek the solution u in the asymptotic form

$$u(x) = \\ = -2\ln(x-a) - \frac{x-a}{a} + \frac{b(x-a)^2}{a^2} + O(x-a)^3 , \quad x \to a , \tag{11.6}$$

where a, b - some fixed real parameters.

The main goal of this paragraph is to find the connection formulas, linking the parameters a, b with the initial data r, s in eq. (11.2).

It was shown previously in [35] that eq. (11.4) may possess a solution with infinite number of poles. The reasoning lies in a non-linear transformation of eq. (11.4) to a system of two equations of the Riccaty type, whose solution is either rational or meromorphic function. The rational solutions are obviously cancelled and so only meromorphic ones remain. It is clear that no coordinates of poles and parameters b are obtained by this method.

We'll advance further in this direction applying the method of isomonodromic deformations, developed above in § 3 for Painlevé III equation. The monodromy data for the corresponding Ψ-function appear to be, as usual, the global invariants for the solution u. If $x \to 0$ and u satisfies the asymptotics (11.2), then the system for Ψ-function, as it was proved in § 8, reduces to Bessel equations and the monodromy matrices are calculated explicitly through the initial data r, s. On the other hand, if we apply the asymptotics (11.6) instead of u in the system for Ψ-function as $x \to a$, we reduce it to the Mathien equation. Thus, taking an appropriate transform, we obtain the Ψ-function as periodic or modified Mathieu function, through which it is possible to express the monodromy data. The combination of the two calculations gives the exact functional equations for the singularity parameters a, b in eq. (11.6):

$$\begin{cases} \arg V(0,a,\nu) = -\frac{1}{2}\arg \dfrac{Ae^{\frac{i\pi r}{4}} + Be^{-\frac{i\pi r}{4}}}{A-B-1}, & (11.7) \\[2ex] \mu(a,\nu) = \frac{r}{4} - \frac{1}{2}, & (11.8) \end{cases}$$

where $\nu^2 = \frac{3}{4}b + \frac{3}{8}$,

$$A = 2^{r} e^{s/2} \Gamma^2\left(\frac{1}{2} + \frac{r}{4}\right), \quad B = 2^{-r} e^{-s/2} \Gamma^2\left(\frac{1}{2} - \frac{r}{4}\right). \qquad (11.9)$$

In eq. (11.7) the function V is the solution of modified Mathieu equation

$$\frac{d^2V}{dz^2} - \left(\nu^2 - \frac{a^2}{2}\,\mathrm{ch}\,2z\right)V = 0$$

with the asymptotics

$$V(z,a,\nu)\longrightarrow e^{-z/2}\exp\left(\frac{ia}{2}e^{z}\right),\quad z\longrightarrow+\infty .$$

The function μ in eq. (11.8) is the Floquet exponent for the solution $V(iz,a,\nu)$ of periodic Mathieu equation.

It seems that eqs. (11.7), (11.8) have no exact solutions for almost any τ, s. However, the asymptotic solutions as $a\to\infty$ can be calculated effectively. For example, the asymptotic distribution of poles at infinity are presented by the following formula

$$a_n=\pi\left(n-\frac{1}{2}\right)-\frac{æ}{2}\ln 8\pi n+\frac{1}{2}\arg\Gamma\left(\frac{1}{2}+iæ\right)+$$
$$+\frac{1}{2}\arg\left(Ae^{\frac{i\pi\tau}{4}}+Be^{-\frac{i\pi\tau}{4}}\right)(A-B)+O\left(\frac{\ln n}{n}\right),\quad n\longrightarrow\infty \tag{11.10}$$

where $æ=\frac{1}{\pi}\ln\left|\frac{A-B}{2\pi}\right|$ and A, B are defined by eq. (11.9). We'll show also that there is no finite point $x\in\mathbb{R}$ of condensation of poles (see Section IV). Combining this result with that of Chapter 8 it is possible to describe the asymptotic behaviour of general real-valued solution of eq. (11.1). Almost all initial data τ, s in eq. (11.2) $(|\tau|<2)$ produces meromorphic solutions with infinite number of poles, tending to infinity with asymptotic distribution (11.10). In only case, when there is the constraint on τ and s:

$$e^{s/2}=2^{-\tau}\,\Gamma\left(\frac{1}{2}-\frac{\tau}{4}\right)\Gamma^{-1}\left(\frac{1}{2}+\frac{\tau}{4}\right), \tag{11.11}$$

the poles vanish and there exists smooth as $x>0$ one-parameter solution, having the asymptotics (11.3), where

$$\alpha=-\frac{2}{\sqrt{\pi}}\sin\frac{\pi\tau}{4} \tag{11.12}$$

(see [5] and also formula (8.32) in Chapter 8).

Finally we give a short sketch of some extra results obtained in [5]. The regular solution (11.3) of PIII equation (11.1) was studied there for all positive values of α. If $\alpha < 2/\sqrt{\pi}$, then the solution is smoothly continued up till the origin, having there the asymptotics (11.2), and the connection formula (11.12) holds. If $|\alpha| = \frac{2}{\sqrt{\pi}}$, then the solution remains smooth (see [5]), and near the origin the asymptotics changes into

$$u(x) = \pm 2 \ln\left[-x\left(\ln\frac{x}{2} + C\right)\right] + O(x^4 \ln^2 x), \quad x \to 0 ,$$

where C is Euler constant. This case corresponds to the limit case $r \to \pm 2$ in eq. (11.2). If, at last, $\alpha > \frac{2}{\sqrt{\pi}}$ the solution becomes meromorfic function with the poles a_n tending to zero ([5]):

$$a_n \sim 4 \exp\left[-\frac{\pi n}{2\mu} - \frac{1}{\mu} \arg \Gamma(i\mu)\right], \; n \to \infty ,$$

where $\alpha = \frac{2}{\sqrt{\pi}} \operatorname{ch} \pi\mu.$

It is quite natural to achieve the latter results by the technique of isomonodromic deformations developed in the present paper. We have no doubt that an asymptotic procedure here is quite similar to those described below. Although we have not yet proved the latter formulae.

I. THE MONODROMY DATA

We remind here briefly the main necessary facts of isomonodromic deformation method, developped in Chapter 3. for Painlevé III equation. Under the change of variables

$$u \longmapsto \pi - iu , \quad x \longmapsto \frac{x}{2}$$

the equation

$$u_{xx} + \frac{1}{x} u_x + \sin u = 0$$

ransforms into eq. (11.1). The system (1.26) is rewritten in the form

$$\frac{d\Psi}{d\lambda} = \left\{ -\frac{ix^2}{4}\sigma_3 - \frac{xu_x}{4\lambda}\sigma_1 + \frac{1}{\lambda^2}(\sigma_2 \operatorname{sh} u - i\sigma_3 \operatorname{ch} u) \right\} \Psi , \qquad (11.13)$$

where as usual $\Psi = \Psi(\lambda, x, u)$ - 2×2 complex-valued matrix, $\sigma_1, \sigma_2, \sigma_3$ - Pauli matrices, $x, u = u(x)$ - real-valued parameters.

Near the irregular points $\lambda = 0$, $\lambda = \infty$ we define two solutions Ψ, Φ of the system (11.13) by their asymptotics:

$$\Psi \longrightarrow \exp\left(-\frac{ix^2}{4}\sigma_3 \lambda\right), \quad \lambda \to \infty , \qquad (11.14)$$

$$\Phi \longrightarrow i\left(\operatorname{sh}\frac{u}{2} - \sigma_1 \operatorname{ch}\frac{u}{2}\right)\exp\left(-\frac{i\sigma_3}{\lambda}\right), \ \lambda \to 0 . \qquad (11.15)$$

The connection matrix Q links together the solutions Ψ and Φ :

$$Q = \Phi^{-1}\Psi , \qquad (11.16)$$

while the Stokes matrices arise for each of the solutions as we circle around the irregular points. Let Ψ_1 (Ψ_2) be the solutions of eq. (11.13) in upper (lower) half-plane λ with asymptotics (11.14) while $\arg \lambda = 0$ ($\arg \lambda = \pi$). The pair of solutions Φ_1, Φ_2 is defined in a similar way using the asymptotics (11.15). Then the equations take place (see Chapter 1)

$$\Psi_1(\lambda) = \Psi_2(\lambda)\begin{pmatrix} 1 & 0 \\ T & 1 \end{pmatrix} ,$$

$$\Phi_1(\lambda) = \Phi_2(\lambda)\begin{pmatrix} 1 & T \\ 0 & 1 \end{pmatrix} , \qquad (11.17)$$

where T is a Stokes multiplier, which is obviously independent of λ. In our case it is also independent of x together with the connection matrix Q.

THEOREM 11.1. The function $u=u(x)$ satisfies Painlevé III equation (11.1) if the connection matrix (11.16) Q and the Stokes multiplier T, defined by eqs. (11.17), do not depend on x.

For the proof see Chapter 3, theorem 3.2 (and also see ref. [16] were it was initially established).

Knowing the monodromy data Q and T we can solve the inverse problem i.e. find the solution Ψ of eq. (11.13) by solving a similar integral equation (see below eq. (11.25)). Then for the derivative u_x of the Painlevé function we have the formula

$$u_x(x) = 2ix \lim_{\lambda\to\infty} \lambda \Psi_{12}(\lambda, x) \exp\left(-\frac{ix^2}{4}\lambda - \frac{i}{\lambda}\right) , \qquad (11.18)$$

which coinsides with formula (1.29) in Chapter 1.

REMARK 11.1. The monodromy data Q and T remain to be the global invariants of the Painlevé function $u(x)$ when x becomes complex-valued (see [16]). The function $w = \exp\frac{u}{2}$ appears to be meromorphic for $|x| > 0$ according to Painlevé property of eg. (11.4). Thus, taking an exponent of the asymptotics (11.6), we obtain the first terms of Laurent series at the pole $x = a$. The parameters a, b standing in its coefficients are uniquely determined by the data Q and T. It means that having Q and T we are able to expand the solution u beyond the pole while passing along the real axis.

There are some constraints on the complex-valued parameters T, Q_{ij}, $i,j=1,2$, which follow from the symmetries of eq. (11.13) and $u(x)$ been real-valued. *)

$$Q_{11}=Q_{22}=-\bar{Q}_{11}\,,\quad Q_{12}=-\bar{Q}_{21}\,, \tag{11.19}$$

$$T=-\bar{T}\,,\quad Q_{11}T=Q_{21}-Q_{12} \tag{11.20}$$

(compare with eqs. (1.58), (1.59) in Chapter 1). The constraints (11.19), (11.20) remain just two arbitrary real parameters in Q and T. We can choose them taking $|Q_{11}|$ and $\arg Q_{12}$, or $|Q_{11}|$ and T.

In fact we have calculated the matrix Q through the initial data (11.2) r, s in Chapter 8(eq (8.27)). It is easy to prove that all our calculations in Chapter 8 remain true after the transform $u\mapsto\pi-iu$, $x\mapsto x/2$, where $|\mathrm{Re}\, r|<2$ in eq. (11.2). Thus we get the result for the connection matrix

$$Q=\frac{1}{2\pi i}\begin{pmatrix} A-B & Ae^{-\frac{i\pi r}{4}}+Be^{\frac{i\pi r}{4}} \\ Ae^{\frac{i\pi r}{4}}+Be^{-\frac{i\pi r}{4}} & A-B \end{pmatrix}, \tag{11.21}$$

---

*) In the notations of Chapter 1 we have that $T=\bar{p}-p$, $Q_{11}=1/\sqrt{1-|p|^2}$, $Q_{12}=p/\sqrt{1-|p|^2}$. The equation (11.19) denote that in present chapter we deal with the part of the general manifold of solutions to P III equation for which $p=-\bar{q}$, $|p|>1$.

where

$$A=2^{\tau}e^{s/2}\Gamma^{2}\left(\frac{1}{2}+\frac{\tau}{4}\right),\quad B=2^{-\tau}e^{-s/2}\Gamma^{2}\left(\frac{1}{2}-\frac{\tau}{4}\right).$$

For the regular near the origin Painlevé function ($u(0)=s$, $u_x(0)=0$, or $\tau=0$ in eq. (11.2)) the matrix Q simplifies

$$Q=-i\begin{pmatrix} sh\frac{s}{2} & ch\frac{s}{2} \\ ch\frac{s}{2} & sh\frac{s}{2} \end{pmatrix}. \tag{11.22}$$

The Stokes multiplier T is determined from eqs. (11.20), (11.21) in the form

$$T=2i\sin\frac{\pi\tau}{4}. \tag{11.23}$$

Finally let us write out an integral equation equivalent to eq. (11.13). Denoting $\Psi=(\Psi_1,\Psi_2)$ the columns of the matrix Ψ, we have

$$\Psi_1(\lambda,s)e^{\theta}=\begin{pmatrix}1\\0\end{pmatrix}+\frac{T}{2\pi i}\int_{-\infty}^{0}\frac{\Psi_2(\xi,x)}{\lambda-\xi}e^{\theta}d\xi+$$

$$+\frac{Q_{21}}{2\pi i\,Q_{22}}\int_{-\infty}^{\infty}\frac{\Psi_2(\xi,x)}{\lambda-\xi}e^{\theta}d\xi, \tag{11.24}$$

where $\theta=\frac{ix^2}{4}\lambda+\frac{i}{\lambda}$, $\pi\leq\arg\lambda<2\pi$ (see [16], eq.(4.30)). One can obtain eq. (11.24) applying the Cauchy formula to the function $\Psi_2 e^{\theta}$ analitic in lower half-plane and by using eqs. (11.17), (11.19), (11.20).

II. THE Ψ-FUNCTION ASYMPTOTIC EXPANSIONS

We will study here the asymptotic expansions of Ψ-function -

the solution of eq. (11.13) under the assumption that $u=u(x)$ has the following asymptotics

$$u(x)=-2\ln\varepsilon-\frac{\varepsilon}{a}-\frac{b\varepsilon^2}{a^2}+O(\varepsilon^2) \quad , \tag{11.25}$$

where $\varepsilon = x-a$, $\varepsilon\to 0$.

Let us find out the leading term of Ψ-function as $\varepsilon\to 0$. We denote $(\psi_1, \psi_2)^T$ the first column of matrix Ψ and introduce the new variables

$$w_1 = \psi_1 + \psi_2 \quad , \quad w_2 = \psi_2 - \psi_1 \quad .$$

then the system (11.13) takes the form

$$\begin{cases} \dfrac{dw_1}{d\lambda} = -\dfrac{xu_x}{4\lambda}\, w_1 + \left(\dfrac{ix^2}{4} + \dfrac{ie^{-u}}{\lambda^2}\right) w_2 \, , \\[2ex] \dfrac{dw_2}{d\lambda} = \left(\dfrac{ix^2}{4} + \dfrac{ie^{u}}{\lambda^2}\right) w_1 + \dfrac{xu_x}{4\lambda}\, w_2 \, . \end{cases} \tag{11.13}$$

Expanding the coefficients in the power series of ε , we have

$$\frac{ix^2}{4} = \frac{ia^2}{4} + \frac{ia\varepsilon}{2} + \frac{i\varepsilon^2}{4} \, ,$$

$$\frac{xu_x}{4} = -\frac{a}{2\varepsilon} - \frac{3}{4} - \frac{\varepsilon(1+2b)}{4a} + O(\varepsilon^2) \, ,$$

$$e^{u} = \frac{2}{\varepsilon^2} - \frac{1}{a\varepsilon} + \frac{1-2b}{2a^2} + O(\varepsilon) \, ,$$

$$e^{-u} = O(\varepsilon^2) \, .$$

Taking the derivative in λ at the first equation (11.13) and expressing w_2 through the second equation, we transform the system (11.13) into the second order scalar equation on w_1 :

$$\frac{d^2 w_1}{d\lambda^2} + \left(\frac{x^4}{16} + \frac{\frac{1}{4} - \nu^2}{\lambda^2} + \frac{1}{\lambda^4}\right) w_1 =$$

$$= O\left(\varepsilon \lambda^{-2} w_1 + \varepsilon^2 \lambda^{-4} \frac{dw_1}{d\lambda}\right) , \qquad (11.26)$$

where $\nu^2 = \frac{3}{4} b + \frac{3}{8}$, and for w_2 the equation holds

$$w_2 = -\frac{4i}{a^2}\left[\frac{dw_1}{d\lambda} - \left(\frac{a}{2\varepsilon\lambda} + \frac{3}{4\lambda}\right) w_1\right] + O(\varepsilon\lambda^{-1}) .$$

Let us define the pair of functions $W, \overline{W}$ as the basis of solutions for the equation

$$\frac{d^2 W}{d\lambda^2} + \left(\frac{x^4}{16} + \frac{\frac{1}{4} - \nu^2}{\lambda^2} + \frac{1}{\lambda^4}\right) W = 0 \qquad (11.27)$$

with the asymptotics at infinity

$$W \to \exp\left(\frac{ix^2}{4}\lambda\right) ,$$

$$\overline{W} \to \exp\left(-\frac{ix^2}{4}\lambda\right) , \quad \lambda \to \infty, \quad \arg\lambda = 0 . \qquad (11.28)$$

Then for the components of the first column we get the asymptotic formulas

$$\psi_1 = \frac{2i}{x^2}\frac{d\overline{W}}{d\lambda} + \left(\frac{3i}{2a^2\lambda} - \frac{i}{\varepsilon a\lambda} + \frac{1}{2}\right)\overline{W} + O(\varepsilon\lambda^{-1} + \varepsilon^2\lambda^{-3}) , \qquad (11.29)$$

$$\psi_2 = -\frac{2i}{x^2}\frac{d\overline{W}}{d\lambda} + \left(-\frac{3i}{2a^2\lambda} + \frac{i}{\varepsilon a\lambda} + \frac{1}{2}\right)\overline{W} + O(\varepsilon\lambda^{-1} + \varepsilon\lambda^{-3}) .$$

It is easy to prove that these components of Ψ-function satisfy the asymptotics (11.14) as $\lambda \to \infty$. The second column of Ψ-function is constructed in the similar way. Finally the leading term in ε for Ψ is obtained in the form

$$\Psi = \begin{pmatrix} \psi_1 & \overline{\psi}_2 \\ \psi_2 & \overline{\psi}_1 \end{pmatrix} , \tag{11.30}$$

where ψ_1 , ψ_2 are defined by eq. (11.29).

Let us proceed now to the matrix Φ - the solution of eq. (11.13) with asymptotics (11.15). It is easy to perform just the same calculations, making the transform

$$\Phi(\lambda) = P\chi(\xi), \quad \xi = \frac{4}{x^2\lambda} , \quad P = i(sh\,\tfrac{u}{2} - \sigma_1 ch\,\tfrac{u}{2}) .$$

Then for the matrix χ we get the system

$$\frac{d\chi}{d\xi} = \left\{-\frac{ix^2}{4}\sigma_3 + \frac{xu_x}{4\xi}\sigma_1 + \frac{1}{\xi^2}(\sigma_2 sh\,u - i\sigma_3 ch\,u)\right\}\chi ,$$

which differs from eq. (11.13) only by the sign in the second term of rhs . The calculations similar to those made up above, lead us to asymptotic expression as $\varepsilon \to 0$

$$\Phi = P\begin{pmatrix} \chi_1 & \overline{\chi}_2 \\ \chi_2 & \overline{\chi}_1 \end{pmatrix} , \tag{11.31}$$

where

$$\chi_1 = \frac{2i}{x^2}\frac{d\overline{W}}{d\xi} - \left(\frac{3i}{2a^2\xi} + \frac{i}{\varepsilon a\xi} - \frac{1}{2}\right)\overline{W} + O(\varepsilon\xi^{-1} + \varepsilon^2\xi^{-3}),$$

(11.32)

$$\chi_2 = \frac{2i}{x^2}\frac{d\overline{W}}{d\xi} - \left(\frac{3i}{2a^2\xi} + \frac{i}{\varepsilon a\xi} + \frac{1}{2}\right)\overline{W} + O(\varepsilon\xi^{-1} + \varepsilon^2\xi^{-3}),$$

Here $W = W(\xi)$ is the solution of eq. (11.27) in variable ξ, where $\xi = 4/x^2\lambda$.

We'll establish now the domains of applicability for the formal asymptotics (11.30), (11.31).

THEOREM 11.2. Let Ψ, Φ - the solutions of eq. (11.13) with asymptotics (11.14), (11.15), and $u = u(x)$ is given by eq. (11.25). Then Ψ and Φ have the leading terms of their asymptotics as $\varepsilon \to 0$, $\arg\lambda = 0$ given by eqs. (11.30), (11.31), where the cancelled terms are estimated as $O(\varepsilon\lambda^{-1} + \varepsilon^2\lambda^{-3})$ for the components of Ψ and $O(\varepsilon\lambda + \varepsilon^2\lambda^3)$ for the components of Φ.

PROOF. The equation (11.26) is equivalent to the following integral equation

$$\overline{w}_1 = \exp\left(-\frac{ix^2}{4}\lambda\right) + \frac{2}{x}\int_\lambda^\infty \sin\frac{x^2}{4}(\lambda - \zeta)\, \times$$

$$\times \left\{\left[p(\zeta) + \varepsilon\zeta^{-2}g(\zeta,\varepsilon)\right]\overline{w}_1(\zeta) + \varepsilon^2\zeta^{-4}f(\zeta,\varepsilon)w_{1x}(\zeta)\right\}d\zeta, \qquad (11.33)$$

where $p(\lambda) = \dfrac{\nu^2 - \frac{1}{4}}{\lambda^2} - \dfrac{1}{\lambda^4}$, $|f|, |g| < const$.

Suppose that $\overline{w}_1 = \overline{W} + \varepsilon M$, where W is the solution of eqs. (11.27), (11.28). Then $\overline{W}$ satisfies eq. (11.33) when $\varepsilon = 0$, so for the remainder M the estimate holds

$$|M| < C(\varepsilon\lambda^{-1} + \varepsilon^2\lambda^{-3}) .$$

The components of Φ are estimated in similar way. The theorem is proved.

III. THE MATHIEU EQUATION FOR THE LEADING TERM OF Ψ-FUNCTION

We'll concentrate now on equation

$$\frac{d^2W}{d\lambda^2} + \left(\frac{a^4}{16} + \frac{1/4-\nu^2}{\lambda^2} + \frac{1}{\lambda^4}\right) W = 0 , \tag{11.34}$$

and study its solutions at the sectors $\Omega'_j = \{\lambda : \pi j \leq \arg\lambda < \pi(j+1)\}$, $j = 0, 1$. It would be proved that the Stokes multiplier T_0, connecting its solutions in the sectors Ω'_j , and the matrix Q_0, connecting them between the irregular points $\lambda = 0$, $\lambda = \infty$, are just the leading terms as $\varepsilon \to 0$ of the corresponding parameters for eq. (11.13). On the other hand the equation (11.34) is simply transformed into the Mathieu equation, so that parameters T_0 and Q_0 are expressed through the Mathien functions.

It is convenient to perform the calculation of the connection matrix Q_0 on the real axis $\arg\lambda = 0$.

The basis of solutions $W, \overline{W}$ of eq. (11.34) is introduced by their asymptotics:

$$W \to \exp\left(\frac{ia^2}{4}\lambda\right) ,$$
$$\overline{W} \to \exp\left(-\frac{ia^2}{4}\lambda\right) , \quad \lambda \to +\infty . \tag{11.35}$$

Note that eq. (11.34) is invariant under the transform

$$W(\lambda) \mapsto \xi^{-1} W(\xi), \quad \xi = \frac{4}{a^2\lambda}, \tag{11.36}$$

so that the pair of functions $\xi^{-1} W(\xi)$ and $\xi^{-1}\overline{W}(\xi)$ also present the basis of eq. (11.34). We may express the old basis through the new one:

$$W(\lambda) = \frac{a}{2}\lambda\left\{p_0 W\left(\frac{4}{a^2\lambda}\right) + q_0 \overline{W}\left(\frac{4}{a^2\lambda}\right)\right\}, \tag{11.37}$$

where $p_0, q_0 \in \mathbb{C}$.

The asymptotics (11.35) together with eq. (11.37) implies the representation of W near the origin:

$$W(\lambda) \to \frac{a}{2}\lambda\left(p_0 e^{i/\lambda} + q_0 e^{-i/\lambda}\right), \quad \lambda \to 0, \quad \arg\lambda = 0. \tag{11.38}$$

Let us define the matrix Q_0, putting

$$Q_0 = \begin{pmatrix} -\overline{p}_0 & -q_0 \\ \overline{q}_0 & p_0 \end{pmatrix}, \tag{11.39}$$

then from eq. (11.37) we obtain

$$\left(\lambda^{-1}\overline{W}(\lambda), \lambda^{-1}W(\lambda)\right) = \left(-\overline{W}\left(\frac{4}{a^2\lambda}\right), W\left(\frac{4}{a^2\lambda}\right)\right)\frac{a}{2} Q_0. \tag{11.40}$$

It means that matrix (11.39) appears to be the connection matrix for eq. (11.34) according to definition given in Chapter 1 (eq. (11.16)).

The following theorem establishes its properties we'll make use of later.

THEOREM 11.3. The matrix Q_0 , defined by eq. (11.39) satisfies the following identities

1) $\det Q_0 = |q_0|^2 - |p_0|^2 = 1$,

2) $p_0 = -\bar{p}_0$,

3) $\arg W(\frac{2}{a}) = \frac{1}{2} \arg \frac{p_0 - 1}{q_0}$.

PROOF. The identity 1) follows directly from the wronskian conservation theorem for eq. (11.34).

If we put $\lambda = 2/a$ in eq. (11.37), then $\bar{W} W^{-1} = (1-p_0) q_0^{-1}$ which implies 3). Taking the absolute value of both sides of the last equation, we get $|1-p_0|^2 = |q_0|^2$. Solving it together with equation 1), we prove thus the identity 2). The theorem is proved.

Let us turn now to the calculation of the Stokes multiplier which can be defined from the identity

$$(W(\lambda e^{-i\pi}), \bar{W}(\bar{\lambda} e^{i\pi})) = (\bar{W}(\bar{\lambda}), W(\lambda)) \cdot \begin{pmatrix} 1 & 0 \\ -T_0 & 1 \end{pmatrix}$$

$$\lambda \in \bar{\Omega}'_0 \; . \tag{11.41}$$

It is convenient here to introduce the new variables

$$\lambda = \frac{2}{a} e^{i\zeta} \; , \quad W(\lambda) = \sqrt{\lambda}\, \vartheta(\zeta) \; . \tag{11.42}$$

The transform (11.42) translates eq. (11.34) into the Mathieu equation

$$\frac{d^2\vartheta}{d\zeta^2} + \left(\nu^2 - \frac{a^2}{2} \cos 2\zeta\right) \vartheta = 0 \; . \tag{11.43}$$

Applying the Floquet theorem to the eq. (11.43) we may represent the function $W(\lambda)$ in the form

$$W(\lambda)=e^{i\zeta/2}\left(e^{i\mu\zeta}Y_1(\zeta)+e^{-i\mu\zeta}Y_2(\zeta)\right), \tag{11.44}$$

where $Y_{1,2}$ are π-periodic functions. From the (11.44) we have that

$$W(\lambda e^{i\pi})-W(\lambda e^{-i\pi})=2i\cos\mu\pi\,W(\lambda).$$

Compare this equality with the (11.41) we conclude that

$$T_0=2i\cos\pi\mu. \tag{11.45}$$

The last formula coinsides with the result of [55] , where the Stokes multipliers were constructed for Mathieu equation.

We are ready now to compare the connection matrix (11.39) with that of (11.16) and the Stokes multiplier (11.41) with that of (11.17).

THEOREM 11.4. Let Q and T be respectively the connection matrix and the Stokes multiplier for eq. (11.13) where $u(x)$ satisfies the asymptotics (11.25). Then the following estimates take place

$$Q-Q_0=O(\varepsilon),\quad T-T_0=O(\varepsilon), \tag{11.46}$$

where Q_0, T_0 are defined for the Mathieu equation (11.34) by eqs. (11.39), (11.45).

PROOF. According to theorem 11.2 there exists a circular domain $\varepsilon^{2/3+\delta}<|\lambda|<\varepsilon^{-2/3-\delta}$, $\delta>0$, where the solution Ψ is presented by eq. (11.30), and Φ is presented by eq. (11.31). Then we can write down the equation $\Psi=\Phi Q$ as follows:

$$\frac{i}{a\lambda}\begin{pmatrix} -\overline{W}(\lambda), -W(\lambda) \\ \overline{W}(\lambda), \ W(\lambda) \end{pmatrix} + \mathcal{O}(\varepsilon) = \left\{ -\frac{i}{2}\begin{pmatrix} -\overline{W}(\xi), \ W(\xi) \\ \overline{W}(\xi), -W(\xi) \end{pmatrix} + \mathcal{O}(\varepsilon) \right\} Q .$$

Expressing the lhs. through the variable ξ by eq. (11.37) and making use of eq. (11.40), we come to estimate we need

$$Q = Q_0 + \mathcal{O}(\varepsilon) \quad ,$$

which proves the first statement of theorem.

In order to derive the second estimate (11.46). Let us turn to the integral equation (11.33). We'll seek the solution at the form

$$\overline{w}_1(\lambda) = \left\{ e^{-\frac{ia^2}{4}\lambda} + \theta_1(\lambda) T(\varepsilon) e^{\frac{ia^2}{4}\lambda} \right\} (1 + o(1)) , |\lambda| \to \infty , \tag{11.47}$$

where $\theta_1(\lambda) = \begin{cases} 0 \ , & 0 \leq \arg\lambda < \pi \ , \\ 1 \ , & \pi \leq \arg\lambda < 2\pi . \end{cases}$

After the substitution of the ansatz (11.47) into eq. (11.33) we compare the terms multiplied by equal powers of λ and equal exponents. On the boundary of the sector Ω_0' the following asymptotics formula is true

$$\int_{\lambda}^{\infty} (1 - e^{i\frac{a^2}{2}(\lambda - \zeta)}) \zeta^{-K-1} d\zeta =$$

$$= \begin{cases} \frac{\lambda^{-K}}{K} (1 + o(1)) \ , & \arg\lambda \to 0 \ , \\ \left[\frac{\lambda^{-K}}{K} + \frac{2\pi i}{K!} \left(\frac{a^2}{2i} \right)^K e^{i\frac{a^2}{2}\lambda} \right] (1 + o(1)) , & \arg\lambda \to \pi . \end{cases} \tag{11.48}$$

It is proved easily applying the integration by parts and deformating the contour of integration so that the residue at the origin gives the nonvanishing term as $\arg\lambda\to\pi$, $|\lambda|\to\infty$.

The last two integrals in eq. (11.33) been expanded using the asymptotics (11.48), appear to be estimated as $O(\varepsilon)$, while $|\lambda|\to\infty$, $0\leqslant\arg\lambda\leqslant\pi$. Thus the Stokes multiplier $T(\varepsilon)$ satisfy an equation

$$T(\varepsilon)=T_0+O(\varepsilon) ,$$

where T_0 is the Stokes multiplier in eq. (11.13) as $\varepsilon=0$. However the solution of eq. (11.33) when $\varepsilon=0$ coinsides with $\overline{W}(\lambda)$, so T_0 has to be the Stokes multiplier for Mathieu equation (11.34). The theorem is proved.

It seems useful to transform eq. (11.34) into modified Mathien equation. This is performed by the change of variables similar to that of eq. (11.42):

$$\lambda=\frac{2}{a}e^{z} , \quad W(\lambda)=\sqrt{\lambda}\,V(z) . \tag{11.49}$$

Thus for $V=V(z,a,\nu)$ we obtain tha modified Mathien equation [56]

$$\frac{d^2V}{dz^2}-(\nu^2-\frac{a^2}{2}\operatorname{ch}2z)V=0 . \tag{11.50}$$

The boundary conditions eqs. (11.35), (11.38) are translated by the transform (11.49) into following asymptotics:

$$V(z)\to e^{-z/2}\exp\left(\frac{ia}{\sqrt{2}}e^{z}\right) , \quad z\to+\infty , \tag{11.51}$$

$$V(z)\to e^{z/2}p_0\exp\left(\frac{ia}{\sqrt{2}}e^{-z}\right)+q_0e^{z/2}\exp\left(-\frac{ia^2}{\sqrt{2}}e^{-z}\right), z\to-\infty, \tag{11.52}$$

The value $\lambda = \frac{2}{a}$ corresponds to $z = 0$, so that $\arg W(\frac{2}{a}) = \arg V(0, a, \nu)$.

THEOREM 11.5. Let $V = V(z, a, \nu)$ be the solution of the modified Mathieu equation (11.50) with the asymptotics (11.51). Then the parameters a, ν satisfy the equations

$$\begin{cases} \arg V(0, a, \nu) = \frac{1}{2} \arg \frac{Q_{11} - 1}{Q_{12}} , & (11.53) \\ 2i \cos \pi\mu = T , & (11.54) \end{cases}$$

where $\mu = \mu(a, \nu)$ is the Floquet exponent for the solution $V(iz, a, \nu)$ of eq. (11.43), and $T, Q = (Q_{ij})$ are respectively the Stokes multiplier and the connection matrix for eq. (11.13).

PROOF follows immediately from the theorem 11.4, eq. (11.45) and the property 3) of the theorem 11.3.

Let us estimate finally the difference between the precise Painlevé function, satisfying eqs. (11.1), (11.2), and its asymptotics (11.6) as $\varepsilon = x - a \to 0$.

THEOREM 11.6. Let $u = u(x)$ be the solution of eqs. (11.1), (11.2) associated with the monodromy parameters T and Q (eqs. (11.23), (11.16)). Then there exist the values a and $b = \frac{4}{3}\nu^2 - \frac{1}{2}$ satisfying eqs. (11.53), (11.54), such that for $u(x)$ the asymptotics (11.6) holds.

PROOF. Let us construct the solution of the Mathieu equation (11.50) with boundary conditions (11.54), (11.52), assuming $T = T_0$, $Q = Q_0$, where T_0, Q_0 are the monodromy parameters (11.39) , (11.45). The existence of such a solution V might be established by application of Gelfand-Levitan-Marchenko equation to initial equation (11.34). Together with V we have found thus the parameters a, ν satisfying functional equations (11.53), (11.54). As it was proved in classical work [1] by Paul Painlevé , there exists a solution $\hat{u}(x)$ of PIII equation (11.1) having asymptotic expansion (11.25) with a and $b = \frac{4}{3}\nu^2 - \frac{1}{2}$. Substituting this solution into the system (11.13) and calculating the asymptotics of Ψ-function as $\varepsilon \to 0$, we conclude, according to theorem 11.4, that co-responding monodromy data $\hat{T}, \hat{Q}$ satisfy the estimates

$$\hat{T} - T = \hat{T} - T = O(\varepsilon) \quad ,$$

$$\hat{Q} - Q = \hat{Q} - Q = O(\varepsilon) \, .$$

On the other hand the isomonodromic condition (theorem 3.2, Chapter 3) implies that $\hat{T}, \hat{Q}$ are independent of x , i.e. of $\varepsilon = x - a$. This yields immediately the equalities

$$\hat{T} = T \quad , \quad \hat{Q} = Q$$

so that the newly constructed solution $\hat{u}(x)$ has the prescribed asymptotics (11.25) with parameters a, ν taken from equations (11.53), (11.54). The theorem is proved.

IV. THE ASYMPTOTIC DISTRIBUTION OF POLES OF PAINLEVÉ III FUNCTION

The functional equations (11.53), (11.54) on the parameters a, ν appear to be transcendent, i.e. there is apparently no explicit solution for any choice of T and Q . In fact for any fixed Floquet exponent μ there is an equation on a and ν involving an infinite chain fraction (Hill's determinant) ([56] , p.143). However we can derive some qualitative conclusions from eqs. (11.53), (11.54) concerning the distribution of poles, as well as to calculate their asymptotics for large a . Alongside with this task it will be proved that $x = \infty$ is the only condensation point of poles of

Painlevé function with initial condition (11.2).

Let us calculate first the asymptotic solution of Mathieu equation (11.50) for large a. It is convenient to rewrite it in the form of Sturm-Liouville equation

$$\frac{d^2V}{dz^2} + a^2\left(\frac{1}{2}\operatorname{ch} 2z - \frac{\nu^2}{a^2}\right)V = 0. \tag{11.58}$$

This equation together with the boundary conditions (11.51), (11.52) represent a problem of one-dimensional scattering of short waves on the potential

$$U(z) = \frac{1}{2}\operatorname{ch} 2z - \frac{\nu^2}{a^2}. \tag{11.59}$$

The asymptotic solution of this problem is a well-known matter. It is given by the classical WKB-method (see, for example, [38]). The solution V has to be sought at the form

$$V(z) \sim \begin{cases} U^{-\frac{1}{4}}(z)\exp\left\{ia\int\limits_{z_0}^{z}\sqrt{U(\zeta)}\,d\zeta\right\}, & z > z_0, \\ U^{-\frac{1}{4}}(z)\left[p_0\exp\left\{ia\int\limits_{z}^{-z_0}\sqrt{U(\zeta)}\,d\zeta\right\} + \right. & \\ \left. + q_0\exp\left\{-ia\int\limits_{z}^{-z_0}\sqrt{U(\zeta)}\,d\zeta\right\}\right], & z < z_0. \end{cases} \tag{11.60}$$

The WKB-solutions (11.60) remains true every where outside the neighbourhoods of turning point, i.e. the zeros of potential $U(z)$. Let us denote

$$\nu^2 = \frac{a^2}{2} + 2a\varkappa \,, \tag{11.61}$$

where the value $\varkappa$ we assume to be finite and independent of a. Then the turning points $-z_0$, z_0 lie closely to each other:

$$z_0^2 = \frac{2\varkappa}{a} + O(a^{-3}), \quad a \to \infty .$$

The scattering data p_0, q_0 in eq. (11.60) are expressed explicitly through the values $\varkappa$ an φ

$$p_0 = -ie^{\pi\varkappa}(1 + O(a^{-1})) \,, \; im\, p_0 < 0 \,, \tag{11.62}$$

$$q_0 = \frac{\sqrt{2\pi}\, exp\left[\frac{\pi\varkappa}{2} + i\varkappa(\ln|\varkappa| - 1) - 2i\varphi\right]}{\Gamma(\frac{1}{2} + i\varkappa)}(1 + O(a^{-1})) \,, \tag{11.63}$$

where

$$\varphi = \frac{a}{2}\int_{z_0}^{\infty}\left\{\sqrt{2ch\, 2z - 1 - 4\varkappa a^{-1}} - e^{z}\right\}dz \,.$$

The formulas (11.62), (11.63) remain true for any $\varkappa \in \mathbb{R}$. They are proved by matching the WKB-solutions (11.60) at the neighbourhood of turning points $-z_0$, z_0, where the solution V is expressed through the parabolic cylinder functions (see the Chapter 10).

The asymptotics of the integral φ as $a \to \infty$ is easily calculated by reducing it to the elliptic integral of the second kind:

$$y = -a + \frac{\varkappa}{2}\left[\ln\frac{2|\varkappa|}{a} - 1 - 4\ln 2\right] + O(a^{-1}\ln a) .$$

Then from eq. (11.62), (11.63) we have

$$\varkappa = \frac{1}{\pi}\ln|p_0| , \quad |p_0| \neq 0 , \tag{11.64}$$

$$\arg q_0 = 2a - 2\pi n + \varkappa \ln a + 3\varkappa \ln 2 - \arg\Gamma\left(\frac{1}{2} + i\varkappa\right) , \tag{11.65}$$

where n is an integer, $n \to \infty$.

It is easy to obtain from (11.65) the asymptotic formula for distribution of poles:

$$a_n = \pi n - \frac{\varkappa}{2}\ln 8\pi n + \frac{1}{2}\arg\Gamma\left(\frac{1}{2} + i\varkappa\right) + \frac{1}{2}\arg q_0 + O\left(\frac{\ln n}{n}\right) , \quad n \to \infty . \tag{11.66}$$

The asymptotics of ν^2 as $n \to \infty$ is derived simply from eq. (11.61) by substitution into it the rhs of eq. (11.65).

Finally we may express $p_0 = Q_{22}$ and $q_0 = -Q_{21}$ through the initial data r, s according to eq. (11.21), and so the formula (11.10) is proved.

REMARK 11.2. Note that there is another way of writing out the asymptotics (11.66). We can exploit instead of eq. (11.63) the asymptotic expression for the phase-shift of the WKB-solution while passing the turning point $-z_0, z_0$. Then the resulting formula (11.66) would comprise the value $T = (q_0 + \bar{q}_0)p_0^{-1}$ instead of $\arg q_0$.

REMARK 11.3. The connection formula (11.62) fails when $\operatorname{Im} p_0$ becomes positive. However this case is reduced to the previous one by a simple transform $u \mapsto -u$ which preserves eq. (11.1). Clearly from eqs. (11.21), (11.39) we get the transformation for monodromy data:

$$p \mapsto -p\,, \quad q \mapsto -q\,.$$

Since the Painlevé equation (11.1) for $-u(x)$ holds the formalism of isomonodromic deformations method, developped in sections II, III, leads again to the Mathieu equation (11.58). The connection formulas (11.62), (11.63) have now the proper signs, so the asymptotic (11.66) is true. The points a_n in (11.66) are now the poles of the inverse function $w = exp\left(-\frac{u}{2}\right)$, which satisfies eq. (11.4).

REMARK 11.4. Let us look at the behaviour of poles in the limit case *) $p_0 \to 0$. Then from eq. (11.62) we conclude that $\varkappa \to -\infty$, so that the coordinate of pole a_n tends to infinity while n is fixed. The equation $p_0 = 0$ is equivalent, according to eq. (11.10), to the condition

$$e^{\frac{s}{2}} = 2^{-\tau}\,\Gamma\left(\frac{1}{2} - \frac{\tau}{4}\right)\Gamma^{-1}\left(\frac{1}{2} + \frac{\tau}{4}\right),$$

which provides the absence of poles and the existence of smooth solution $u = u(x)$ of eq. (11.1) for $x > 0$ (see [5]). Thus the limit case $p_0 \to 0$ leads to "deportation" of poles towards infinity.

THEOREM 11.7. The poles of Painlevé function $w = exp\,\frac{u}{2}$ with the initial condition (11.2) have the asymptotic distribution (11.66) as $x \to \infty$. No other point $x \in \mathbb{R}$ could be the limit point of the set of poles of this function.

---

*) In the terms of monodromy data $p, q = -\bar{p}$ this correspond the case $|p| \to \infty$, $Im\,p = 0$.

PROOF. It is necessary to prove only the second statement of the theorem. Let there exist a sequence of poles $a_n \longrightarrow a^* < \infty$ .We'll consider then the spectrum points ν_n of the periodic Mathieu equation (11.42) at the fixed value of the Floquet exponent μ $(T = 2i\cos\pi\mu)$. While $a = a_n$ are bounded the values of ν_n , satisfying the equation $\mu = \mu(a_n, \nu_n)$, have to go to infinity as $n \to \infty$ ([56], p.163). Then for the modified Mathien equation (11.50)

$$\frac{d^2V}{dz} - \nu^2\left(\frac{a^2}{2\nu^2}\operatorname{ch}2z - 1\right)V = 0$$

we can write out again the WKB-solutions (11.60), where a is replaced by ν and the potential $U(z) = \frac{a^2}{2\nu^2}\operatorname{ch}2z - 1$ as $\nu \to \infty$. It can be shown in a similar way, that the connection formulas (11.62), (11.63) remain true also in this case, when the turning points lie far from each other:

$$z_0^2 \simeq \frac{2\varkappa}{a} \;, \quad \varkappa \simeq \nu^2 \;,$$

according to (11.61).

Thus the equation (11.62) does not hold as $\nu_n \to \infty$ because the value $|\rho_0| = \exp\pi\varkappa$ remains fixed and finite. This contradiction proves the theorem.

Chapter 12. LARGE-TIME ASYMPTOTICS OF THE SOLUTION OF THE CAUCHY PROBLEM FOR MKdV EQUATION

In this chapter we apply the scheme of isomonodromic deformation method (IDM) to asymptotic analysis of nonlinear evolutionary equations integrable by the inverse scattering transform (IST) method. We concentrate on the Cauchy problem for modified Korteweg- de Vries equation as a typical example of exactly solvable equations demonstrating typical large-time asymptotic behaviour of its solutions.

The Cauchy problem we consider here reads

$$\begin{cases} y_t - 6y^2 y_x + y_{xxx} = 0 \\ y\big|_{t=0} = y_0 \in \mathcal{S}_{\mathbb{R}}(-\infty, \infty), \end{cases} \tag{12.1}$$

where $\mathcal{S}_{\mathbb{R}}(-\infty,\infty)$ is Schwarz's space of real-valued functions defined on the real axis $\mathbb{R}^1$.

Within the framework of IST the equation

$$y_t - 6y^2 y_x + y_{xxx} = 0 \tag{12.2}$$

is associated with self-conjugate Dirac operator on the real axis

$$L(y) = i\sigma_3 \frac{d}{dx} + y\cdot\sigma_1 \ . \tag{12.3}$$

Let us fix the parameters of the Cauchy problem (12.1) in terms of the scattering data for the operator $L(y_0)$. Consider $\mathbb{F}(x,z)$ and $\widetilde{\mathbb{F}}(x,z)$ to be the standard Jost solutions of the equation

$$L(y_0)\Psi = z\Psi \ , \tag{12.4}$$

defined by the conditions

$$\mathbb{F}(x,z) \sim \exp\{-i\sigma_3 z x\}, \quad x \to +\infty,$$

$$\tilde{\mathbb{F}}(x,z) \sim \exp\{-i\sigma_3 z x\}, \quad x \to -\infty, \quad z \in \mathbb{R}.$$

The transition matrix

$$\mathbb{T}(z) = [\tilde{\mathbb{F}}(z)]^{-1} \mathbb{F}(z) \tag{12.5}$$

would be written down of the form

$$\mathbb{T}(z) = \begin{pmatrix} a(z) & c(z) \\ b(z) & d(z) \end{pmatrix}, \quad z \in \mathbb{R}.$$

The well-known algebraic constraints take place for the functions a, b, c, d (see, for example, [43])

$$d(z) = \bar{a}(z)$$

$$c(z) = \bar{b}(z) \Leftrightarrow \sigma_1 \mathbb{T}(z) \sigma_1 = \bar{\mathbb{T}}(z), \tag{12.6}$$

$$|a(z)|^2 - |b(z)|^2 = 1 \Leftrightarrow \det \mathbb{T}(z) = 1, \tag{12.7}$$

$$a(-z) = \bar{a}(z)$$

$$b(-z) = -\bar{b}(z) \Leftrightarrow \mathbb{T}^T(-z) = \mathbb{T}^{-1}(z). \tag{12.8}$$

Besides we have the following asymptotic properties of these functions:

a) $b(z) \in \mathbb{S}_{\mathbb{C}}(\mathbb{R}^1)$,

where $\mathbb{S}_{\mathbb{C}}(\mathbb{R}^1)$ is Schwarz's space of complex-valued functions on the real axis.

b) $a(z) = A_+(z) = \lim_{\varepsilon \to 0} A(z + i\varepsilon)$,

$$A(z) = \exp\left\{ \frac{1}{2\pi i} \int_{-\infty}^{\infty} \frac{\ln(1 + |b(\xi)|^2)}{\xi - z} d\xi \right\},$$

(12.9)

$$\operatorname{Im} z \neq 0 \ .$$

The scattering data for the operator $L(y_\circ)$ together with the initial data of the Cauchy problem (12.1) would be parametrized, following an usual scheme, through the reflection coefficient

$$r(z) = b(z)/a(z) \ .$$

It is easy to derive from (12.6) - (12.9) the basic properties of $r(z)$:

1. $r(z) \in \mathbb{S}_{\mathbb{C}}(\mathbb{R}^1)$

2. $r(-z) = -\overline{r(z)} \Rightarrow \bar{r}(0) = -r(0)$,

3. $1 - |r(z)|^2 = \dfrac{1}{|a(z)|^2} \Rightarrow |r(z)| < 1$ for all $z \in \mathbb{R}$.

4. $A(z) = \exp\left\{ -\dfrac{1}{2\pi i} \displaystyle\int_{-\infty}^{\infty} \dfrac{\ln(1 - |r(\xi)|^2)}{\xi - z} d\xi \right\}$, $\operatorname{Im} z \neq 0$,

$$a(z) = A_+(z) \ , \quad z \in \mathbb{R} \ .$$

The basic aim of this Chapter is to give an explicit description of asymptotics as $t \to +\infty$ of the solution $y(t,x)$ (12.1) in terms of the function $\tau(z)$. The central role in a proposed method play the results, obtained in the main text of the paper, dealing with asymptotic parametrization of solutions to PII equation by the monodromy data. More precisely, we need here the information about real-valued solution $u(\eta)$ of the equation

$$u_{\eta\eta} - \eta u - 2u^3 = 0 . \tag{12.10}$$

According to the notations of Chapter 7 they belong to the submanifold $\mathring{\mathbb{M}}_{\mathbb{R}}$, i.e. their monodromy data p, q satisfy the equalities

$$-p = q = i\rho , \quad -1<\rho<1 . \tag{12.11}$$

Let us denote these solutions by $u(\eta|\rho)$ and remind their properties, following from the results of Chapter 7:

1. The asymptotics of $u(\eta|\rho)$ take place as $|\eta|\to\infty$:

$$u(\eta|\rho)=\frac{\rho}{2\sqrt{\pi}}\,\eta^{-1/4} e^{-2/3\,\eta^{3/2}}(1+o(1)) , \quad \eta \to +\infty , \tag{12.12}$$

$$u(\eta|\rho)=\sqrt{2}\,\alpha\cdot(-\eta)^{-1/4}\cos\Big\{\frac{2}{3}(-\eta)^{3/2} -$$

$$-\frac{3}{2}\alpha^2\ln(-\eta)+\varphi\Big\}+o\big((-\eta)^{-1/4}\big) , \quad \eta\to-\infty , \tag{12.13}$$

where

$$\alpha>0 , \quad \alpha^2=-\frac{1}{2\pi}\ln(1-\rho^2) ,$$

$$\varphi=-3\alpha^2\ln 2-\arg\Gamma(-i\alpha^2)+\frac{\pi}{2}\,\mathrm{sign}\,\rho-\frac{\pi}{4} .$$

2. The canonical solutions $\Psi_5(\lambda, \eta|\rho)$ and $\Psi_2(\lambda, \eta|\rho)$ of the system (1.9), constructed through the parameters $u \equiv u(\eta|\rho)$ and $v \equiv u_\eta(\eta|\rho)$ $(x \equiv \eta)$, provide a solution of the following matrix Riemann-Hilbert (RH) problem

$$\chi_-(\lambda) = \chi_+(\lambda)\begin{pmatrix} 1 & i\rho e^{-\frac{8}{3}i\lambda^3 - 2i\eta\lambda} \\ i\rho e^{\frac{8}{3}i\lambda^3 + 2i\eta\lambda} & 1-\rho^2 \end{pmatrix}, \mathrm{Im}\,\lambda = 0, \tag{12.14}$$

$$\chi(\lambda) \to I, \quad \lambda \to \infty, \quad \mathrm{Im}\,\lambda \neq 0. \tag{12.15}$$

where

$$\chi(\lambda) = \begin{cases} \Psi_2(\lambda, \eta|\rho)\exp\{\frac{4}{3}i\lambda^3\sigma_3 + i\eta\lambda\sigma_3\}, & \mathrm{Im}\,\lambda > 0, \\ \Psi_5(\lambda, \eta|\rho)\exp\{\frac{4}{3}i\lambda^3\sigma_3 + i\eta\lambda\sigma_3\}, & \mathrm{Im}\,\lambda < 0. \end{cases} \tag{12.16}$$

The function $u(\eta|\rho)$ itself may be expressed through a solution $\chi(\lambda, \eta|\rho)$ of RH problem (12.15), (12.14) by the constraints

$$\frac{\partial\Psi}{\partial\eta}\Psi^{-1} = -i\lambda\sigma_3 - u\sigma_2, \tag{12.17}$$

$$\Psi(\lambda, \eta|\rho) = \chi(\lambda, \eta|\rho)\exp\{-\frac{4}{3}i\lambda^3\sigma_3 - i\eta\lambda\sigma_3\}.$$

Return now to the Cauchy problem (12.1). We separate the halfplane of (x, t) variables, following the work [43], into three characteristic domains:

$$\Omega_+ = \{(x,t) : \varepsilon t \leq x \leq Ct; \quad 0 < \varepsilon < C < \infty\},$$

$$\Omega_- = \{ (x,t) : -Ct \leq x \leq -\varepsilon t , \quad 0 < \varepsilon < C < \infty \} ,$$

$$\Omega_0 = \{ (x,t) : -Ct^{1/3} \leq x \leq Ct^{1/3} , \ 0 < C < \infty \} .$$

The following theorem produces the complete asymptotic description of the leading term as $t \to +\infty$ of solution to the Cauchy problem (12.1). The proof of it constitutes the major part of the Chapter.

THEOREM 12.1. Let the initial data $y_0(x)$ of the Cauchy problem (12.1) correspond to the reflection coefficient $r(z)$. Then the following statements are true:

1. The solution $y(x,t)$ decreases as $t \to +\infty$ faster than any power of t in the domain Ω_+ . The leading term of asymptotics may be expressed through the integral

$$y(x,t) = \frac{i}{\pi} \int_{-\infty}^{\infty} r(\xi) e^{i8t\xi^3 + 2ix\xi} d\xi (1 + o(1)) ,$$

or, if we assume that $r(z)$ is analytic in the upper half-plane,

$$y(x,t) \approx \frac{i}{2\sqrt{3\pi t}} r\left(\frac{i}{2}\sqrt{\frac{x}{3t}}\right) \cdot \left(\frac{x}{3t}\right)^{-1/4} e^{-2t \cdot \left(\frac{x}{3t}\right)^{3/2}} .$$

2. In the domain Ω_0 the solution $y(x,t)$ is self-similar and is represented by PII -function:

$$y(x,t) = (3t)^{-1/3} \cdot u\left(\frac{x}{(3t)^{1/3}} \,\Big|\, ir(0)\right) (1 + o(1)) \quad t \to +\infty .$$

3. If $(x,t) \in \Omega$ - the solution $y(x,t)$ oscillates. The leading term of asymptotics has the form

$$y(x,t)=\frac{1}{\sqrt{3 z_0 t}}\,\alpha\cos\{2t(-\frac{x}{3t})^{3/2}-\alpha^2\ln 12t+\varphi\}+o(t^{-1/2})$$

$$t \to +\infty$$

where

$$z_0 = \frac{1}{2}\sqrt{-\frac{x}{3t}}$$

$$\alpha^2 \equiv \alpha^2(z_0) = -\frac{1}{2\pi}\ln\left(1-|r(z_0)|^2\right)$$

$$\varphi \equiv \varphi(z_0) = -3\alpha^2 \ln z_0 - 4\alpha^2 \ln 2 - \arg\Gamma(-i\alpha^2) -$$

$$-\frac{\pi}{4} - \arg r(z_0) - \frac{2}{\pi}\int_{-z_0}^{z_0} \ln|\xi - z_0|\, d\ln|a(\xi)| .$$

REMARK 12.1. As it follows from (12.12), (12.13) the asymptotic formulae from N 2 and N 1 (N 3) go one into another under the assumption

$$|x/t| \to 0 \ , \quad x/t^{+1/3} \to \infty\ (-\infty) \ .$$

Then the Painlevé function

$$u(\eta|\rho)\ , \quad \eta = x/(3t)^{1/3}\ , \quad \rho = ir(0)$$

links the asymptotic behaviour of the solution $y(x,t)$ in the domains Ω_+ and Ω_- for which the domain Ω_0 is a transitional zone.

Remark 12.2. The asymptotic formula derived in N 1 was first obtained by A.B.Shabat [45] . The most interesting oscillating part of the asymptotics (N 2) for the case of KdV, NS and SG was first calculated by V.E.Zakharov and S.V.Manakov [45] . Finally, a complete description of the asymptotic solution to the Cauchy problem (12.1) given in Theorem 12.1 was proposed by M.J.Ablowitz and H.Segur [43] , [44] .

PROOF of the theorem 12.1. The inverse scattering problem is equivalent to a certain matrix RH problem. We remind , according to [29] the corresponding procedure of the reduction. Denote

$$\Psi_+(z) = \left(\frac{f(z)}{a(z)}, \tilde{g}(z) \right),$$

$$\Psi_-(z) = \left(\tilde{f}(z), \frac{g(z)}{\bar{a}(z)} \right),$$

where f, g, $\tilde{f}$, $\tilde{g}$ are the columns of Jost matrices

$$\mathbb{F} = (f, g), \quad \tilde{\mathbb{F}} = (\tilde{f}, \tilde{g}).$$

The well-known asymptotic properties of functions $\mathbb{F}$, $\tilde{\mathbb{F}}$ yield the following statements:

a) $\Psi_+(z)$ is analytic as $\operatorname{Im} z > 0$ and

$$\Psi_+(z) = (I + o(1)) \exp\{-izx\sigma_3\}, \; z \to \infty, \; \operatorname{Im} z \geq 0.$$

b) $\Psi_-(z)$ is analytic as $\operatorname{Im} z < 0$ and

$$\Psi_-(z) = (I + o(1)) \exp\{-izx\sigma_3\}, \; z \to \infty, \; \operatorname{Im} z \leq 0.$$

In terms of $\Psi_\pm$ functions the constraint (12.5) reads

c)
$$\Psi_-(z) = \Psi_+(z)\begin{pmatrix} 1 & \bar{r}(z) \\ -r(z) & 1-|r(z)|^2 \end{pmatrix}.$$

Introducing the functions $\chi_\pm$

$$\chi_\pm(z) = \Psi_\pm(z)\exp\{izx\sigma_3\},$$

we rewrite a) - c) in the form of RH problem:

$$\chi_-(z) = \chi_+(z)\begin{pmatrix} 1 & \bar{r}(z)e^{-2izx} \\ -r(z)e^{2izx}, & 1-|r(z)|^2 \end{pmatrix}, \quad \operatorname{Im} z = 0,$$

$$\chi(z) \to I, \quad z \to \infty .$$

This problem is exactly equivalent to initial inverse scattering problem for the $L(y_0)$ operator.

The dynamics of the scattering data in t along the solutions of MKdV equation is a well-known matter (see, for example, [43]):

$$r(z) \longmapsto r(z)\, e^{8iz^3 t} .$$

Hence we come to an asymptotic analysis as $t \to \infty$ of the following RH problem:

$$\chi_-(z)=\chi_+(z)\begin{pmatrix} 1 & \bar{r}(z)e^{-2izx-8itz^3} \\ -r(z)e^{2izx+8itz^3}, & 1-|r(z)|^2 \end{pmatrix}, \tag{12.18}$$

$$\mathrm{Im}\, z = 0 \ ,$$

$$\chi(z)\to I \ , \quad z\to\infty . \tag{12.19}$$

Clearly, it is equivalent to the system of singular integral equations of the form

$$\chi_+(z) =$$

$$= I+\frac{1}{2\pi i}\int_{-\infty}^{\infty}\frac{1}{\xi-z-i0}\chi_+(\xi)\begin{pmatrix} 0 & , -\bar{r}(\xi)e^{-it\theta(\xi)} \\ r(\xi)e^{it\theta(\xi)}, & |r(\xi)|^2 \end{pmatrix} d\xi \tag{12.20}$$

where $\theta(z)=2z\frac{x}{t}+8z^3$.

The solution $y(x,t)$ (12.1) is reconstructed through the function $\chi(\lambda, x, t)$, satisfying (12.18), (12.19) by means of inversion formulae

$$\frac{d\Psi}{dx}\Psi^{-1} = -iz\sigma_3 - y\cdot\sigma_2 \ , \tag{12.21}$$

which is equivalent to

$$y=\frac{i}{\pi}\int_{-\infty}^{\infty}\chi_{22}^{+}(\xi)r(\xi)e^{it\theta(\xi)}d\xi . \tag{12.22}$$

We are going to investigate the system (12.20) closely following the scheme of Chapter 4, where we have studied an "isomonodromic" RH problem (4.8) . Note that the technical approaches we use here, are quite similar to those applied first by S.V.Manakov in [63] while calculating the asymptotics of solution to nonlinear Schrödinger equation.

1. THE DOMAIN Ω_+ . The situation here is quite simple. No stationary point lie on the contour of integration in (12.20) because

$$\theta'(z) = 2\frac{x}{t} + 24z^2 \geqslant 2\varepsilon > 0 \ , \quad \forall z \in \mathbb{R} \, .$$

Hence, splitting the system (12.20) into two scalar integral equations on functions χ^+_{21}, χ^+_{22} and applying the procedure of asymptotic simplification developped in Chapter 4, we come to the estimate

$$\chi^+_{22}(z) = 1 + o(1), \ t \to \infty \, . \tag{12.23}$$

A remainder term in (12.23) decreases faster than any power of t as $t \longrightarrow \infty$ uniformly in $z \in \mathbb{R}$, and it has no oscillating exponent. Substituting (12.23) into (12.22) and taking into account the remarks made above we obtain the asymptotic representation for $y(x,t)$ declared in theorem 12.1:

$$y(x,t) = \frac{i}{\pi}\int_{-\infty}^{\infty} r(\xi) e^{it\theta(\xi)} d\xi \ (1 + o(1)) \, . \tag{12.24}$$

If we assume for simplicity that $r(z)$ is analytic in the upper half-plane, then, evaluating an integral in (12.24) by the steepest descent method, we get finally

$$y(x,t)=\frac{i}{2\sqrt{3\pi}}\, r\left(\frac{i}{2}\sqrt{\frac{x}{3t}}\right)\cdot\left(\frac{x}{3t}\right)^{-1/4} t^{-1/2} e^{-2t\cdot(x/3t)^{3/2}}\left(1+o(1)\right),$$

$$t\longrightarrow+\infty\ ,\quad (x,t)\in\Omega_+\ .$$

2. THE DOMAIN Ω_0 . The most simple way to prove the second statement of the theorem proceeds as follows. First rescale the variable z :

$$z\longmapsto\lambda = z\cdot(3t)^{1/3}\ . \tag{12.25}$$

In a new variable λ the RH problem (12.18) - (12.19) may be rewritten in the form

$$\chi_-(\lambda)=\chi_+(\lambda)\begin{pmatrix} 1 & ,\ \bar r\left(\frac{\lambda}{(3t)^{1/3}}\right)e^{-2i\lambda\eta-\frac{8}{3}i\lambda^3} \\ -r\left(\frac{\lambda}{(3t)^{1/3}}\right)e^{2i\lambda\eta+\frac{8}{3}i\lambda^3} & ,\ 1-\left|r\left(\frac{\lambda}{(3t)^{1/3}}\right)\right|^2 \end{pmatrix}, \tag{12.26}$$

$$\chi(\lambda)\longrightarrow I\ ,\quad \lambda\longrightarrow\infty,$$

where $\eta=x/(3t)^{1/3}$.

The variable η remains bounded under the conditions $(x,t)\in\Omega_0$

$$|\eta|<C\ ,\quad t\longrightarrow+\infty\ . \tag{12.27}$$

Alongside with it we have

$$r\left(\frac{\lambda}{(3t)^{1/3}}\right)\longrightarrow r(0),\quad \forall\lambda\in\mathbb{R}\ ,\ t\to\infty\ . \tag{12.28}$$

It means that the Riemann-Hilbert problem (12.26) is transformed as

$t \longrightarrow \infty$ into an "isomonodromic" one (12.15) with the parameter

$$\rho = i r(0) . \qquad (12.29)$$

The latter satisfies the inequality (12.11) because of the properties 2, 3 of the reflection coefficient $r(z)$. In terms of MKdV solutions the transformation of the Riemann-Hilbert problem (12.26) as $t \to \infty$ into (12.11) together with renormalization (12.25) of a spectral parameter implies the representation

$$y(x,t) \sim (3t)^{-1/3} u(\eta|\rho), \ t \to +\infty ,$$

$$(x,t) \in \Omega_0 \ , \ \eta = \frac{x}{(3t)^{1/3}} \ , \ \rho = i r(0) . \qquad (12.30)$$

We are going to omit standard through cumbersome analysis of singular integral equations which transforms the formal reasoning above into the strict proof, justifying the estimate (12.30).

3. THE DOMAIN Ω_- . This domain presents the most labourous part of our analysis. The difficulties arise while the stationary phase points

$$z = \pm z_0 \ , \quad z_0 = \frac{1}{2}\sqrt{-\frac{x}{3t}}$$

appear on the contour of integration as $\frac{x}{t} \leq -\varepsilon < 0$. The situation here is quite similar to that of Chapter 4, where we have failed to extract in elementary manner the asymptotics to PII solution directly from the integral equations.

Here again we make use of some standard "isomonodromic" RH problem (12.15), through the parameters ρ and η turn to be connected with the reflection coefficient $r(z)$ by more complicated constraints. Namely, it would be shown, that under conditions

$$t \to +\infty, \quad z_0 = \frac{1}{2}\sqrt{-\frac{x}{3t}} \geqslant \delta > 0 , \tag{12.31}$$

the initial RH problem (12.18) transforms into (12.15), where

$$\rho = -|r(z_0)| ,$$

$$\eta = \frac{x + x_0}{(3t)^{1/3}} , \tag{12.32}$$

$$2z_0 \cdot x_0 \equiv \frac{2}{\pi}\int_{-z_0}^{z_0} \ln|\xi - z_0| d\ln|a(\xi)| + \arg r(z_0) - \frac{\pi}{2} .$$

It is clear that again the solution $y(x,t)$ (12.1) might be approximated through the Painlevé transcendent $u(\eta|\rho)$ in the domain Ω_- . There are however two significant differences from the case of domain Ω_0 . In the first place, the variable η defined by (12.32), tends to $-\infty$ as $t \to +\infty$. Hence an asymptotics

$$y(x,t) \sim (3t)^{-1/3} u(\eta|\rho)$$

transforms in such a way, that $u(\eta|\rho)$ substitutes on its asymptotics (12.13). In the second place, the asymptotic parameters depend on a slow variable x/t .

We proceed now to a proof and concretization of the declared above connection of RH problems (12.18) and (12.15).

The matrix integral equation (12.20) is equivalent to two disconnected with one another systems of scalar equations:

$$\chi^+_{11}(z) = 1 + \frac{1}{2\pi i}\int_{-\infty}^{\infty} \frac{1}{\xi - z - i0}\, r(\xi)\, \chi^+_{12}(\xi) e^{it\theta(\xi)} d\xi ,$$

$$\chi^{+}_{12}(z) = -\frac{1}{2\pi i}\int_{-\infty}^{\infty} \frac{1}{\xi - z - i0}\,\bar{r}(\xi)\chi^{+}_{11}(\xi)e^{-it\theta(\xi)}d\xi + \tag{12.20a}$$

$$+\frac{1}{2\pi i}\int_{-\infty}^{\infty}\frac{1}{\xi - z - i0}\,|r(\xi)|^{2}\chi^{+}_{12}(\xi)d\xi\ ;$$

$$\chi^{+}_{21}(z) = \frac{1}{2\pi i}\int_{-\infty}^{\infty}\frac{1}{\xi - z - i0}\,r(\xi)\chi^{+}_{22}(\xi)e^{it\theta(\xi)}d\xi\ ,$$

$$\chi^{+}_{22}(z) = 1 - \frac{1}{2\pi i}\int_{-\infty}^{\infty}\frac{1}{\xi - z - i0}\,\bar{r}(\xi)\chi^{+}_{21}(\xi)e^{-it\theta(\xi)}d\xi + \tag{12.20b}$$

$$+\frac{1}{2\pi i}\int_{-\infty}^{\infty}\frac{1}{\xi - z - i0}\,|r(\xi)|^{2}\chi^{+}_{22}(\xi)\,d\xi\ .$$

Under the condition (12.31) we investigate the systems (12.20a) and (12.20b) as $t \to \infty$. Consider first the system (12.20b). Starting from an asymptotics (just similar to that of (4.18) in Chapter 4)

$$\frac{1}{2\pi i}\int_{-\infty}^{\infty}\frac{f(\xi)e^{i\sigma t\theta(\xi)}}{\xi - z \mp i0}\,d\xi =$$

$$= \pm f(z)e^{i\sigma t\theta(\lambda)}\,\eta(\sigma(\pm z_0^2 \mp z^2)) + O\left(\frac{1}{\sqrt{t}(z^2 - z_0^2)}\right),$$

$$t \to +\infty\ ,\quad \sigma = \pm 1\ , \tag{12.33}$$

and acting just like as we did in Chapter 4 ignoring the singularities at $z = \pm z_0$, one could reduce the system (12.20b) to the scalar equation

$$\chi^{+}_{22}(z) \approx 1 + \frac{1}{2\pi i} \int_{-z_0}^{z_0} \frac{1}{\xi - z - i0} |r(\xi)|^2 \chi^{+}_{22}(\xi)\, d\xi \, . \tag{12.34}$$

The equation (12.34) is equivalent to a scalar RH problem of the form

$$\chi^{-}_{22}(z) = \chi^{+}_{22}(z)\left(1 - |r(z)|^2 \eta(z_0^2 - z^2)\right),$$

$$\chi_{22}(z) \to 1 \quad , \quad z \to \infty \quad ,$$

the solution of which is given by the formula *)

$$\chi_{22}(z) \approx \exp\left\{-\frac{1}{2\pi i} \int_{-z_0}^{z_0} \frac{\ln(1-|r(\xi)|^2)}{\xi - z} d\xi\right\} =$$

$$= \exp\left\{\frac{1}{\pi i} \int_{-z_0}^{z_0} \frac{\ln |a(\xi)|}{\xi - z} d\xi\right\} \equiv \alpha^{-1}(z) \, . \tag{12.35}$$

Applying now the first equation (12.20b) we have for the solution $\chi_{21}(z)$ a representation

----

*) At this stage all our reasoning have an exactly formal shape. The aim is to obtain some reasonable hypothesis upon the leading term of solution to RH problem (12.18) . Therefore we are just ignore the behaviour of solutions (12.34) at the singular points $\pm z_0$ and a correctness of the equation itself in the neighbourhood of those points.

$$\chi_{21}(z) \approx \frac{1}{2\pi i} \int_{-\infty}^{\infty} \frac{1}{\xi - z} r(\xi) \alpha_{+}^{-1}(\xi) e^{it\theta(\xi)} d\xi . \tag{12.36}$$

The similar analysis of the system (12.20a) yields

$$\chi_{11}(z) \approx \alpha(z) , \tag{12.37}$$

$$\chi_{12}(z) \approx -\frac{A(z)}{2\pi i} \int_{-\infty}^{\infty} \frac{\bar{r}(\xi)\alpha_{+}(\xi)\bar{a}(\xi)e^{-it\theta(\xi)}}{\xi - z} d\xi , \tag{12.38}$$

Therefore, as a result of natural-lloking, through not strict asymptotic analysis of singular integral equation (12.20), we have a piecewise-analytic matrix-valued function $\tilde{\chi}(z)$. Its elements $\tilde{\chi}_{ij}(z)$ are defined by right-hand sides of (12.35) - (12.38). We are going to prove now that $\tilde{\chi}(z)$ solves approximately the initial RH problem in such a way that

$$\tilde{\chi}^{+} G [\tilde{\chi}^{-}]^{-1} = I + O\left(\frac{1}{\sqrt{t}(z^2 - z_0^2)}\right) , \tag{12.39}$$

where $G(z)$ is the conjugation matrix in RH problem (12.18).

Note that because of explicit formulae defining $\tilde{\chi}$, one could check the equation (12.39) directly. The only fact more we need consists of a proof of asymptotics (12.33) for the densities, entering the Cachy integrals (12.36) and (12.38). Those densities contain a boundary value of the function $\alpha(z)$, which has the form

$$\delta_{\pm}(z) = \left(\frac{z - z_0}{z + z_0}\right)_{\pm}^{i\nu} \exp\left\{-i\gamma(z) \pm \eta(z_0^2 - z^2) \ln\left|\frac{a(z_0)}{a(z)}\right|\right\} ,$$

where

$$\nu = \frac{1}{\pi} \ln|a(z_0)| > 0 ,$$

$$\gamma(z) = \frac{1}{\pi} \int_{-z_0}^{z_0} \ln|\xi - z| \, d \ln|a(\xi)| , \qquad (12.40)$$

$\left(\frac{z - z_0}{z + z_0}\right)^{i\nu}$ is the single-valued branch, defined by its jump line, which coinsides with interval $[-z_0, z_0]$.

The behaviour of $\gamma(z)$ in the neighbourhoods of points $\pm z_0$ is described by the constraints (C_0 and C_1 are constants)

$$\gamma(z) - \gamma(\pm z_0) =$$

$$= C_0 (z \mp z_0) \ln|z \mp z_0| + C_1 (z \mp z_0) + O((z \mp z_0)^2 \ln|z \mp z_0|) .$$

Outwards of the points $\pm z_0$ the integral $\gamma(z)$ represents a smooth function of z . As a result we need to establish (12.33) estimates in the case of $f(\xi)$, having the form

$$f(\xi) = F(\xi)(\xi \mp z_0)^{i\nu} ,$$

where

1. $F(z)$ is smooth as $z \neq \pm z_0$,
2. $F(z) - F(\pm z) =$

$$= C_0 (z \mp z_0) \ln|z \mp z_0| + C_1 (z \mp z_0) + O((z \mp z_0)^2 \ln^2 (z \mp z_0)). \qquad (12.41)$$

The standard proof of estimates (12.33) demands an asymptotic analysis by means of the stationary phase method of the following integrals

$$I_{\pm}(F,\nu;t)=\int_{\pm z_0}^{\infty} F(\xi)(\xi \mp z_0)_{+}^{i\nu}\, e^{-it\theta(\xi)}\, d\xi ,$$

where $Im\,\nu = 0$ and $F(\xi)$ satisfies (12.41). All the information we need about asymptotic behaviour of integrals $I_{\pm}(F,\nu;t)$ is given by the following lemma, which could be derived, for example, from [52] :

LEMMA 12.1. The asymptotic expansion takes place as $t\to\infty$

$$I_{\pm}(F,\nu;t)=$$

$$=\frac{1}{2}F(\pm z_0)\,\frac{e^{\pm 16itz_0^3 \pm \frac{\pi\nu}{2} \mp \frac{i\pi}{4}}}{(24z_0t)^{\frac{i\nu}{2}+\frac{1}{2}}}\,\Gamma\left(\frac{i\nu}{2}+\frac{1}{2}\right)+O(t^{-1}\ln^2 t).$$

The lemma 12.1 shows in particular, that similarly to the common stationary phase method, the leading term in t has the form

$$I_{\pm}(F,\nu;t)=\varkappa_{\pm}(t)\,t^{-1/2}\,F(\pm z_0) , \qquad (12.42)$$

where $\varkappa_{\pm}(t)$ are oscillating in t functions, defined through an universal procedure of stationary phase method. Clearly this result is quite sufficient for the proof of estimates (12.33) in a class of functions $f(\xi)$ considered above.

As far as we have proved the asymptotics (12.33) it is easy to check directly the constraints (12.39) to be true. We are going to omit those elementary calculations.

The estimate (12.39) is not an exact characteristic feature of the approximation $\widetilde{\chi}$ to the solution of RH problem (12.18). One can always change the function $\tilde{\chi}$ in such a way that (12.39) remains true. For example, it is possible to modify $\tilde{\chi}$ into χ^m :

$$\tilde{\chi} \mapsto \chi^m(z) = \hat{\chi}^m(z) \begin{pmatrix} \alpha(z) & 0 \\ 0 & \alpha^{-1}(z) \end{pmatrix} ,$$

where

$$\hat{\chi}^m_{11} = \hat{\chi}^m_{22} = 1 ,$$

$$\hat{\chi}^m_{12}(z) = \begin{cases} -\frac{1}{2\pi i}\int_{-\infty}^{\infty} \frac{1}{\xi - z} |a(\xi)|^2 \bar{r}(\xi)\alpha_+^2(\xi) e^{-it\theta(\xi)} d\xi , \ \mathrm{Im}\, z > 0 , \\ -\frac{1}{2\pi i}\int_{-\infty}^{\infty} \frac{1}{\xi - z} \bar{r}(\xi)\left[\alpha_+(\xi)\alpha_-(\xi)\right] e^{-it\theta(\xi)} d\xi , \ \mathrm{Im}\, z < 0 \end{cases} \tag{12.43}$$

$$\hat{\chi}^m_{21}(z) = \begin{cases} \frac{1}{2\pi i}\int_{-\infty}^{\infty} \frac{1}{\xi - z} r(\xi)\alpha_+^{-2}(\xi) e^{it\theta(\xi)} d\xi , \ \mathrm{Im}\, z > 0 , \\ \frac{1}{2\pi i}\int_{-\infty}^{\infty} \frac{1}{\xi - z} r(\xi)\alpha_+^{-1}(\xi)\alpha_-^{-1}(\xi) e^{it\theta(\xi)} d\xi , \ \mathrm{Im}\, z < 0 . \end{cases}$$

Evidently, the equations

$$\left[\chi^m(z)\right]^+ = \chi^{\pm}(z)$$

hold with a remainder term $O\left(\frac{1}{\sqrt{t}(z^2 - z_0^2)}\right)$. Hence, the function $\chi^m(z)$ would solve the RH problem (12.18) in the same sense and with the same precision as the function $\tilde{\chi}$. Note the significant difference between $\hat{\chi}$ and χ^m lying in the fact of absense of any exterior multiplier depending of z before the Cauchy integrals (12.43), which define matrix elements of χ^m . This property of χ^m appears to be very essential on a final stage of the proof.

Thus we have constructed an approximate solution $\chi^{m}(z)$ of the RH problem (12.18). The similar approximation has been used by S.V. Manakov [63] in the case of nonlinear Schrödinger equation while calculating an asymptotics of module to its solution as $t\to\infty$. However the function $\chi^{m}(z,x,t)$ does not produce a leading term of asymptotics of the solution $y(x,t)$. The reason is that in a neighbourhood of stationary phase points $\pm z_0$ the function χ^{m} does not satisfy RH equation. The matrix elements of χ^{m} contain a singularity nonvanishing in t as $z\longrightarrow \pm z_0$. Our next task would be a detailed study of the singularity.

The asymptotic behaviour of function $\chi^{m}(z)$ in the neighbourhood of the points $\pm z_0$, is determined by a behaviour of the integrals

$$\overset{1}{K}_{\pm}(z|f,\tau)=\frac{1}{2\pi i}\int_{-\infty}^{\infty}\frac{1}{\xi-z}\,f(\xi)\alpha_{+}^{\pm 2}(\xi)e^{\mp it\theta(\xi)}\,d\xi\ ,$$

$$\overset{2}{K}_{\pm}(z|f,\tau)=\frac{1}{2\pi i}\int_{-\infty}^{\infty}\frac{1}{\xi-z}\,f(\xi)\left[\alpha_{+}(\xi)\alpha_{-}(\xi)\right]^{\pm 1}e^{\mp it\theta(\xi)}\,d\xi\ ,$$

where $f(\xi)$ is a smooth function. Consider for a certainty the first integral $\overset{1}{K}_{+}(z|f,\tau)$ and transform it as follows

$$\overset{1}{K}_{+}(z|f,\tau)=\frac{F(z_0)}{2\pi i}\int_{-\infty}^{\infty}\frac{1}{\xi-z}e^{-it\theta(\xi)}\left(\frac{\xi-z_0}{\xi+z_0}\right)_{+}^{2i\nu}d\xi\ +$$

$$+\frac{1}{2\pi i}\int_{-\infty}^{\infty}\frac{F(\xi)-F(z_0)}{\xi-z}e^{-it\theta(\xi)}\left(\frac{\xi-z_0}{\xi+z_0}\right)_{+}^{2i\nu}d\xi\ ,$$

where

$$F(\xi)=f(\xi)\exp\{-2i\gamma(\xi)+\eta(z_0^2-\xi^2)\ln\left|\frac{a(z_0)}{a(\xi)}\right|^2\}\ .$$

Thus we have reduced the question to the study of asymptotic behaviour of the latter integrals. They have the form

$$I_0(z) = \int_{-\infty}^{\infty} \frac{1}{\xi - z} \left(\frac{\xi - z_0}{\xi + z_0} \right)^{2i\nu} e^{-it\theta(\xi)} d\xi$$

and

$$I(\tilde{F}|z) = \int_{-\infty}^{\infty} \frac{\tilde{F}(\xi)}{\xi - z} e^{-it\theta(\xi)} (\xi - z_0)_+^{2i\nu} d\xi ,$$

where $\tilde{F}(\xi)$ satisfies the conditions (12.41) together with the auxiliary equation $\tilde{F}(z_0)=0$. Simplifying the latter integral further in the neighbourhood of the point z_0 , we come to a model integral of the form

$$I_1(z) = \int_{-\infty}^{\infty} \frac{\ln\xi \cdot \xi_+^{2i\nu} e^{-it\xi^2}}{\xi - \delta} d\xi , \quad \delta = z - z_0 .$$

Applying the same refereces as above for the lemma 12.1 one can derive estimate

$$I_1^{\pm}(z) = O\left(\delta \ln\delta - \frac{1}{\sqrt{t}} \ln\delta - \frac{1}{\sqrt{t}} \ln t\right), \ t \to \infty , \ 0 < \delta < 1 .$$

Differing from the integral $I_1(z)$ another one $I_0(z)$ has a nonvanishing singularity in the neighbourhood of z_0 as $t \to \infty$. Although it remains bounded as $\delta \to 0$ its precise asymptotics is not so simple to derive (see [71]). However as it always occurs in similar cases, there is no need of the detailed study of those integrals. Through the transform

$$K_{+}^{1}(z\,|\,f,\tau) =$$

$$= \frac{1}{2\pi i}\, F(z_0)\, I_0(z) + O\Big((z-z_0)\ln(z-z_0) + \frac{1}{\sqrt{t}}\ln|z-z_0| + \frac{\ln t}{\sqrt{t}}\Big). \quad (12.44)$$

we have localized the singularity by retaining it in the <u>universal</u> integral $I_0(z)$. The representations similar to that of (12.44) take place for all the remainder integrals $K_{\pm}^{j}(z|f,\tau)$ $(j=1,2)$ in the neighbourhoods of both singular points $\pm z_0$. We have thus proved the following proposition, which would be crucical for further calculations.

PROPOSITION 12.1. The integrals $K_{+}^{j}(z|f,z)$ and $K_{+}^{j}(z|\tilde{f},\tilde{\tau})$ coinside in the leading term as $t \to +\infty$ in the neighbourhood of the points $\pm z_0$ under the condition

$$f(\pm z_0)e^{-2i\gamma(\pm z_0)} = \tilde{f}(\pm z_0)e^{-2i\tilde{\gamma}(\pm z_0)}.$$

the similar condition for the coinsidence of leading terms of $K_{-}^{j}(z\,|\,f,\tau)$ and $K_{-}^{j}(z|f,\tau)$ reads

$$f(\pm z_0)e^{2i\gamma(\pm z_0)} = \tilde{f}(\pm z_0)e^{2i\tilde{\gamma}(\pm z_0)}.$$

The result states in the Proposition 12.1 leads us directly to the basic problem - the reduction of initial RH problem (12.18) to the "isomonodromic" one (12.15) , (12.32). It arises naturally, as we have seen while proving the theorem 12.1.

Let us introduce alongside with the function $\chi^{m}(z)$ a new function $\mathring{\chi}^{m}(z)$ through the formulae (12.43) where the tramsform has been made

$$r(z)\mapsto r_0(z)=i|r(z_0)|\exp\left\{i\frac{z}{z_0}\left[2\gamma(z_0)+\arg r(z_0)-\frac{\pi}{2}\right]\right\} .$$

We demonstrate now that $\overset{\circ}{\chi}{}^m(z)$ provides the same asymptotic behaviour as $\chi^m(z)$ at the singular point z_0 . More precisely, the following estimate takes place

$$[\overset{\circ}{\chi}{}^m(z)]^{-1}\chi^m(z)=\exp[-i\sigma_3\gamma(\pm z_0)]+$$

$$+O\left[(z\mp z_0)\ln|z\mp z_0|+\frac{1}{\sqrt{t}}\ln|z\mp z_0|+\frac{1}{\sqrt{t}}\ln t\right] \tag{12.45}$$

$$z\sim\pm z_0 .$$

In fact the equation

$$|r_0(z)|=|r(z_0)|$$

yields

$$|a_0(z)|\equiv|a_0(z_0)|=|a(z_0)| ,$$

$$\alpha_0(z)=\left(\frac{z-z_0}{z+z_0}\right)^{i\nu}, \tag{12.46}$$

$$\gamma_0(z)\equiv 0 .$$

On the other hand, due to the property 2 of the reflection coefficient $r(z)$ and an obvious equality $\overline{\gamma}(z)=\gamma(z)=-\gamma(-z)$, the equations hold

$$r(\pm z_0)e^{2i\gamma(\pm z_0)}=r_0(\pm z_0) ,$$

$$\overline{r}(\pm z_0)e^{-2i\gamma(\pm z_0)}=\overline{r}_0(\pm z_0) . \tag{12.47}$$

Thus the estimate (12.45) is a corollary of the Proposition 12.1.

Consider now the matrix

$$G_0(z)=\begin{pmatrix} 1 & \bar{r}_0(z)e^{-it\theta(z)} \\ -r_0(z)e^{it\theta(z)} & 1-|r_0(z)|^2 \end{pmatrix}$$

and show that the function

$$\hat{\chi}(z)=\left[\overset{\circ}{\chi}{}^{m}(z)\right]^{-1}\chi^{m}(z)$$

provides an "asymptotic undressing" of conjugation matrix $G(z)$ in initial RH problem (12.18), so that it transforms into $G_0(z)$.

a) THE NEIGHBOURHOOD OF z_0. Since the estimate (12.45) holds the following equalities take place

$$\hat{\chi}^{+}(z)\,G(z)\left[\hat{\chi}^{-}(z)\right]^{-1}=$$

$$=\left[\exp(-i\sigma_3\gamma(z_0))+\omega(z)\right]\left\{G(z_0)+O(z-z_0)\right\}\left[\exp(i\sigma_3\gamma(z_0))+\tilde{\omega}(z)\right]=$$

$$=e^{-i\sigma_3\gamma(z_0)}\,G_0(z)\,e^{i\sigma_3\gamma(z_0)}+\omega'(z)=$$

$$=G_0(z)+O\left[(z-z_0)\ln|z-z_0|+\frac{1}{\sqrt{t}}\ln|z-z_0|+\frac{1}{\sqrt{t}}\ln t\right]$$

Here we have put

$$\omega(z),\ \tilde{\omega}(z),\ \omega'(z)=O\left[(z-z_0)\ln|z-z_0|+\frac{1}{\sqrt{t}}\ln|z-z_0|+\frac{1}{\sqrt{t}}\ln t\right].$$

b) THE NEIGHBOURHOOD OF $-z_0$. The very similar calculation as above proves the equality

$$\hat{\chi}^+(z)\,G(z)\left[\hat{\chi}^-(z)\right]^{-1} =$$

$$= G_0(z) + O\left[(z+z_0)\ln|z+z_0| + \frac{1}{\sqrt{t}}\ln|z+z_0| + \frac{1}{\sqrt{t}}\ln t\right] , \quad z \sim -z_0 .$$

c) BEYOND THE NEIGHBOURHOODS OF $\pm z_0$. For the function $\overset{\circ}{\chi}{}^{m}(z)$ the estimate holds

$$\overset{\circ}{\chi}{}^{m}_{+}(z)\,G_0(z)\left[\overset{\circ}{\chi}{}^{m}_{-}(z)\right]^{-1} = I + O\left(\frac{1}{\sqrt{t}\,(z^2 - z_0^2)}\right).$$

Hence, outwards the points $\pm z_0$ we have

$$\hat{\chi}^+(z)\,G(z)\left[\hat{\chi}^-(z)\right]^{-1} = G_0(z) + O\left(\frac{1}{\sqrt{t}\,(z^2 - z_0^2)}\right).$$

Bringing together all the results obtained above at a), b) and c) we establish the following proposition

PROPOSITION 12.2. There exists a piecewise analytic function $\hat{\chi}(z)$, such that

1. $\hat{\chi}(z)$ is analytic as $\operatorname{Im} z > 0$ and as $\operatorname{Im} z < 0$.
2. If $z \to \infty$ and $\operatorname{Im} z \neq 0$, then

$$\hat{\chi}(z) = I + O\left(\frac{1}{z}\right).$$

3. If $\operatorname{Im} z = 0$, $t \to +\infty$ and $-\dfrac{x}{12t} = z_0^2 > 0$, then

$$\hat{\chi}^+(z)\,G(z)\left[\hat{\chi}^-(z)\right]^{-1} = G_0(z) + R(x,t,z) ,$$

where

$$R(x,t,z)=\begin{cases} O\left(\frac{1}{\sqrt{t}(z^2-z_0^2)}\right), & |z^2-z_0^2|\geq t^{-\varepsilon} \\ O\left[(z^2-z_0^2)\ln|z^2-z_0^2|+\frac{1}{\sqrt{t}}\ln|z^2-z_0^2|+\frac{1}{\sqrt{t}}\ln t\right], & |z^2-z_0^2|<t^{-\varepsilon}, \end{cases}$$

$$0<\varepsilon<\frac{1}{2}.$$

Consider now the RH problem of the form (12.18), (12.19), but with $G_0(z)$ as a conjugation matrix:

$$\overset{\circ}{X}_-(z)=\overset{\circ}{X}_+(z)G_0(z), \qquad (12.48)$$

$$\overset{\circ}{X}(z)\longrightarrow I, \quad z\longrightarrow\infty, \quad \mathrm{Im}\, z\neq 0.$$

Note that

$$r_0(z)e^{it\theta(z)}=i|r(z_0)|\exp\{2iz(x+x_0)+8itz^3\},$$

where

$$x_0=\frac{2\gamma(z_0)+\arg r(z_0)-\frac{\pi}{2}}{2z_0}.$$

Hence, making the familiar rescaling of variables

$$z\mapsto\lambda=z\cdot(3t)^{1/3} \qquad (12.49)$$

and introducing a parameter

$$\eta=\frac{x+x_0}{(3t)^{1/3}} \qquad (12.50)$$

we reduce the RH problem (12.48) to "isomonodromic" problem (12.15) where the parameters ρ and η have been choosen according to

(12.32). Thus our aim declared at the beginning of the third stage of the proof, has been reached.

Assume $y_0(x,t)$ to be a solution of MKdV equation associated with the RH problem (12.48). It provides the equation on Ψ-function (we ignore here the dependence of x through $r(z_0)$)

$$\frac{\partial \overset{\circ}{\Psi}}{\partial x}\overset{\circ}{\Psi}{}^{-1} = -i\sigma_3 z - y_0 \sigma_2 \, , \tag{12.51}$$

where $\overset{\circ}{\Psi}(z) = \overset{\circ}{\chi}(z)\exp\left\{-4iz^3 t\sigma_3 - izx\sigma_3\right\}$.

Comparing it with that of (12.17) and taking into account the constraints (12.49), (12.50) between the parameters λ, z, x and η, one concludes that

$$y_0(x,t) = (3t)^{-\frac{1}{3}} u(\eta|\rho) \, , \tag{12.52}$$

where ρ and η are defined through (12.32). As a final step we substitute instead of the Painlevé function $u(\eta|\rho)$ its asymptotics (12.13) and rewrite the leading term through the old variables

$$y_0(x,t) =$$

$$= \frac{1}{\sqrt{3 z_0 t}} \cdot d \cdot \cos\left\{16 t z_0^3 - d^2 \ln 12 t - 3d^2 \ln z_0 - 4d^2 \ln 2 - \right.$$

$$\left. - \arg\Gamma(-id^2) - \frac{\pi}{4} - 2\gamma(z_0) - \arg r(z_0)\right\} + o(t^{-\frac{1}{2}}) \, ,$$

where

$$z_0 = \frac{1}{2}\sqrt{-\frac{x}{3t}} \quad ,$$

$$\alpha^2 = -\frac{1}{2\pi}\ln(1-|r(z_0)|^2),$$

$$\gamma(z_0)=\frac{1}{\pi}\int_{-z_0}^{z_0}\ln|\xi-z_0|\,d\ln|a(\xi)|.$$

All we need for a completion of the proof is to justify the following asymptotics for the solution $y(x,t)$ of the Cauchy problem

$$y(x,t)=y_0(x,t)+o(t^{-1/2}),$$

$$t\to+\infty,\quad -\frac{x}{12t}=z^2>0. \tag{12.53}$$

Here the transition from $\hat{\chi}$ to the new function χ^m we have made above will play its role. The absense of multipliers depending on z before the Cauchy integrals which define the matrix elements $\hat{\chi}^m_{ik}$, means that the leading terms in $t\to+\infty$ of functions $\hat{\chi}^m(z)$ and $\hat{\mathring{\chi}}^m(z)$ coinside. As a matter of fact, consider for example $\hat{\chi}^m_{12}(z)$ and $\hat{\mathring{\chi}}^m_{12}(z)$ as $\operatorname{Im} z>0$. According to (12.42) the asymptotics take place as $t\to+\infty$

$$\hat{\chi}^m_{12}(z)=$$

$$=-\frac{1}{2\pi i}\frac{1}{z_0-z}\cdot\varkappa_+(t)t^{-1/2}|a(z_0)|^2\bar{r}(z_0)e^{-2i\gamma(z_0)-it\theta(z_0)}+$$

$$-\frac{1}{2\pi i}\frac{1}{z_0+z}\cdot\varkappa_-(t)t^{-1/2}|a(-z_0)|^2\bar{r}(-z_0)e^{-2i\gamma(-z_0)-it\theta(z_0)}+O\left(\frac{\ln^2 t}{t}\right),$$

$$\hat{\mathring{\chi}}^m_{12}(z)=$$

$$= \frac{1}{2\pi i} \cdot \frac{1}{z_0 - z} \cdot \varkappa_+(t) t^{-1/2} |a_0(z_0)|^2 \bar{r}_0(z_0) e^{-it\theta(z_0)} +$$

$$- \frac{1}{2\pi i} \cdot \frac{1}{z_0 + z} \cdot \varkappa_-(t) \cdot t^{-1/2} |a_0(-z_0)|^2 \bar{r}_0(-z_0) e^{-it\theta(z_0)} + O\left(\frac{\ln^2 t}{t}\right) .$$

Taking into account the constraints (12.46), (12.47) one obtains

$$\hat{\chi}^m_{12}(z) - \hat{\mathring{\chi}}^m_{12}(z) = O\left(\frac{\ln^2 t}{t}\right) .$$

In other words the function

$$\hat{\chi}(z) = [\mathring{\chi}^m]^{-1} \chi^m =$$

$$= e^{-\sigma_3 \ln d_0(z)} [\hat{\mathring{\chi}}^m]^{-1} \hat{\chi}^m e^{\sigma_3 \ln d(z)}$$

satisfies the following estimate as $|\operatorname{Im} z| \geq \varepsilon$

$$\hat{\chi}(z) = \exp\{-\sigma_3 \ln d_0(z)\} \left\{ I + O\left(\frac{\ln^2 t}{t}\right)\right\} \exp\{\sigma_3 \ln d(z)\}, \qquad (12.54)$$

$$t \to +\infty .$$

That is the function $\hat{\chi}$, linking together the RH problems (12.18) and (12.15), is in some sense very close to the unit matrix.

We are ready now to establish the estimate (12.53).

$$\chi_{as}(z) = \mathring{\chi}(z) \hat{\chi}(z)$$

asymptotic solution of the RH problem (12.18). that even after "switching on" an auxiliary dependence of x through $r(z_0) = r\left(\frac{1}{2}\sqrt{\frac{x}{3t}}\right)$ the equation (12.51) remains true in a leading order of t . The remainder term is of order $O(\ln^2 t / t^{3/2})$

(12.51) and (12.54) we have

$$\frac{\partial \Psi_{as}}{dx}\Psi_{as}^{-1} = -iz\sigma_3 - y_0\sigma_2 + o(t^{-1/2})$$

uniformly in z , $|\operatorname{Im} z| \geqslant \varepsilon$. Here as usual we have denoted

$$\Psi_{as} = \chi_{as}\cdot \exp\{-ixz\sigma_3 - 4itz^3\sigma_3\}.$$

Therefore in order to justify the asymptotics (12.53) it is sufficient to prove that the precise solution $\chi(z,x,t)$ of the RH problem (12.18), (12.19) is related to that of χ_{as} by a constraint

$$\chi = (I + o(t^{-1/2}))\chi_{as} \qquad (12.55)$$

uniformly in z , $|\operatorname{Im} z| \geqslant \varepsilon$.

For that purpose consider an expression

$$D(z) \equiv \chi(z)\chi_{as}^{-1}(z).$$

The function $D(z)$ is analytic in the upper and lower half-planes of z , and it tends to I as $z \to \infty$, $\operatorname{Im} z \neq 0$. On the real axis we have

$$D_- = D_+ + g\cdot D_+ ,$$

where $g(x,t,z)$ posesses the same properties as the function $R(x,t,z)$ from the Proposition 12.2. Therefore

$$D(z) = I - \frac{1}{2\pi i}\int_{-\infty}^{\infty}\frac{1}{\xi - z}\, g(\xi)D_+(\xi)\,d\xi =$$

$$= I + \int_{-\infty}^{\infty}\frac{\omega(\xi,x,t)}{\xi - z}\,d\xi ,$$

where ω is a matrix-valued function with the same properties as R. Moreover its matrix elements $\omega_{ij}(z)$ conatin obviously an oscillating exponents of the form $exp\{\pm it\cdot\theta(\xi)\}$. This two properties of ω yield the estimate

$$D(z)=I+o(t^{-1/2}), \quad Im z \neq 0 ,$$

which proves (12.55).

Chapter 13. THE DYNAMICS OF ELECTROMAGNETIC IMPULSE IN A LONG LASER AMPLIFIER

We consider the Maxwell-Bloch system of integro-differential equations, describing the propagation of coherent electromagnetic impulse in a "two-level media" [58]

$$\begin{cases} E_t + E_x = 2\pi i \int_{-\infty}^{\infty} n(\omega)\, u(\omega) \bar{v}(\omega) d\omega \\ u_t = i\omega u + iEv \\ v_t = -i\omega v + i\bar{E}u, \quad x>0, t>0. \end{cases} \tag{13.1}$$

Here $E = E(x,t)$ is the complex-valued amplitude of electric field, u, v - the probability amplitudes for the "two-level atoms" being stayed respectively at the upper (lower) positions. The function $n = n(\omega)$ is a given function, such that

$$N = \int_{-\infty}^{\infty} n(\omega) d\omega > 0 \quad ,$$

which corresponds to the case of inversely populated media.

The initial impulse being exited at the entrance of the amplifier is described by the boundary value conditions

$$\begin{aligned} & E(0,t) = E_0(t) \quad \text{as} \quad t>0, \\ & E(x,0) = 0 \; ; \\ & u(\omega,x,t) = 1, \; v(\omega,x,t) = 0 \quad \text{as} \quad t = 0. \end{aligned} \tag{13.2}$$

Furthermore we assume that E is real-valued and the boundary function E_0 has the asymptotics

$$E_0(t)=ct^{\nu}+O(t^{\nu+1}),\ \nu>0\ ,\ t\to 0\ . \tag{13.3}$$

We proceed now with a calculation of asymptotics of the solution E as $x\to+\infty$. It describes the form of the resulting impulse at along distance in the laser amplifier.

The boundary value problem for the system (13.1) is completely integrable (see [17]) in such a way that its two last equations play the role of L-operator in the Lax pair whereas ω is a spectral parameter. The reflection coefficient $R(\omega,x)$, defined in a standard manner, is presented by the formula ([17])

$$R(\omega,x)=$$

$$=R(\omega,0)\exp 2ix\left(\omega-\frac{\pi}{2}\int_{-\infty}^{\infty}\frac{n(\omega')}{\omega'-\omega-i0}d\omega'\right), \tag{13.4}$$

where $R(\omega,0)$ is defined through the boundary condition (13.2). Never the less the direct calculation of the asymptotics at a large x of the potential E seems to be difficult.

The extremely effective way of asymptotic integration to (13.1), (13.2) was developed in ref. [17] . It is based on an observation of the "quasi-self-similar" behaviour of solution E at large x . We are now going to give a short review of this method, which essentially, exploits the asymptotic solutions of PIII equation.

Due to the hyperbolic property of the system (13.1), i.e. the finite velocity of impulse propagation, the solution E is equal to zero as $t<x$. For a large x in the domain $t>x$ the system (13.1) is reduced to a single equation

$$U_{zz} + \frac{1}{z} U_z = \sin U \tag{13.5}$$

by the change of variables

$$U = \int_0^t E(x,\tau)d\tau, \quad z = 4\Omega_0\sqrt{x(t-x)}, \tag{13.6}$$

where $\Omega_0^2 = \frac{\pi}{2}N$. The reason for the latter transform is just follows. It appears that $|E|$ is of order $\mathcal{O}(x)$ as $t \sim x + \mathcal{O}(\sqrt{x})$ so that two last equations in (13.1) admit the WKB solution as $\omega \sim x^{-1}$

$$u \sim \cos\int_0^t E(x,\tau)d\tau, \quad v \sim i\sin\int_0^t E(x,\tau)d\tau.$$

On substituting them into the first equation we arrive to the sine-Gordon equation

$$U_{xt} + U_{tt} = 4\Omega_0^2 \sin U , \tag{13.7}$$

which in its turn reduces to PIII equation (13.5).

It remains now to determine the initial conditions for the equation (13.5). Clearly there is a domain near the wave front $t = x$, where E remains small because $E = 0$ as $t < x$. Thus we may linearise the system (13.1) in this domain:

$$U_{xt} + U_{tt} = 4\Omega_0^2 U . \tag{13.8}$$

The boundary conditions (13.2) are now

$$U(x,0) = 0 , \quad U_t(0,t) = E_0(t) . \tag{13.9}$$

It turns so that eq. (13.8) is exactly the linearization of sine-Gordon equation (13.7), thus the solution of it would serve as a linear limit for the Painlevé function $U = U(z)$ from (13.5). Since U is small we may substitute eq. (13.5) by its linearization - the Bessel equation

$$U_{zz} + \frac{1}{z} U_z = U .$$

We select its solution growing at infinity

$$U = U_0 \, I(z) = U_0 \frac{e^z}{\sqrt{2\pi z}} \left(1 + O(z^{-1})\right), \quad z \to \infty . \tag{13.10}$$

On the other hand we may solve explicitly the linear boundary problem (13.8), (13.9) and thus calculate its asymptotics as $x \to \infty$, $0 < z < \ln^{\rho} x$, $\rho > 0$. The leading term may be obtained directly from the integral representation of the solution U by applying the stationary phase method:

$$U = c \cdot \left(\frac{z}{x}\right)^{\nu+1} (8\Omega_0^2)^{-\nu-1} \Gamma(\nu+1) \frac{e^z}{\sqrt{2\pi z}} \left(1 + O(z^{-1})\right), \tag{13.11}$$

$$z \to \infty ,$$

where c is the coefficient in asymptotics (11.3) for the boundary function $E_0(t)$. It is clear from (13.11) that U remains small as $z < \nu \ln z$.

We conclude that the two asymptotic solutions (13.11) and (13.10) are matched together since $z \to \infty$, $U \ll 1$ and the constant U_0 in (13.10) satisfies the matching condition

$$U_0 = c\, \Gamma(\nu+1) (8\Omega_0^2)^{-\nu-1} \left(\frac{z}{x}\right)^{\nu+1} . \tag{13.12}$$

We have found thus the initial condition for the PIII equation (13.5) *)

$$U(0)=U_0 \qquad U_z(0)=0 \quad , \tag{13.13}$$

where U_0 is given by the formula (13.12). Note that the expression (13.12) depends on the second, "slow" with respect to z, variable $s = z/x$. Therefore the solution of eq. (13.7)

$$U=U(z, U_0(s))$$

was called quasi-self-similar in ref [17].

Returning to the initial function $E=U_t$, we obtain from (13.6)

$$E=\frac{8\Omega^2 x}{z} U_z(z, U_0(\tfrac{z}{x})) \quad , \tag{13.14}$$

where U is the solution of PIII equation (13.5) with the initial condition (13.13).

The connection formulae found in Section 8 provide the effective asymptotics for E as $z \to \infty, (\nu \ln x < z < x^{1-\varepsilon}, \ \varepsilon > 0)$. The equation (13.5) is easily reduced to that of (3.5) by the transform $U=\pi + u$. We may able now to apply directly the asymptotic formulae (8.8), (8.30):

$$E=-\alpha \frac{8\Omega_0^2 x}{z^{3/2}} \sin\left(z-\frac{\alpha^2}{16}\ln z + \varphi\right)(1+o(1)) \quad , \tag{13.15}$$

---

*) The second equality arises because here we must consider only the regular solutions to eq. (15.5).

where

$$\alpha^2 = -\frac{16}{\pi} \ln \sin \frac{U_0(s)}{2} ,$$

$$\varphi = \frac{\pi}{4} + \arg \Gamma\left(\frac{i\alpha^2}{16}\right) - \frac{\alpha^2}{8} \ln 2 , \quad s = \frac{z}{x} , \tag{13.16}$$

and the $U_0(s)$ function is defined by (13.12).

As a concluding remark note that the method of isomonodromic deformations combined with the method of inverse scattering problem provide an opportunity to prove the formal asymptotics constructed above. We describe briefly the way it is proved.

The basic step is the calculation of the reflection coefficient $\hat{R}(\omega, x)$ corresponding to the solution (13.14). We make use of function

$$\Psi = \begin{pmatrix} u & -\bar{v} \\ v & \bar{u} \end{pmatrix} \tag{13.17}$$

which is precisely the Ψ-function, satisfying the system (1.26) with $w = U_z$, $u \mapsto U - \pi$, $x \mapsto z$. Due to the fact that U is the solution of PIII equation (13.5) the Ψ-function (13.17) satisfies also the last two equations (13.1), where

$$E = 8\Omega_0^2 \frac{x}{z} U_z ,$$

$$\lambda = \frac{\omega z}{8x\Omega_0^2} , \quad z = 4\Omega_0 \sqrt{x(t-x)} . \tag{13.18}$$

As we have seen above the solution (13.14) is true in the domain $0 < z < x^{1-\varepsilon}$, thus the leading term of the scattering matrix $\hat{S}$ for the potential E is calculated as

$$\hat{S} = \Psi^{-1}\Big|_{z=0}\Psi\Big|_{z\sim x} .$$

Since the asymptotics for the matrix Ψ has been constructed in Chapter 8 we easily get the expression for the reflection coefficient $\hat{R} = \hat{S}_{12}/\hat{S}_{22}$

$$\hat{R}(\omega,s,x) = \hat{R}(\omega,s,0)\exp\{2ix(\omega - \Omega_0^2\omega^{-1})\} , \tag{13.19}$$

where $\hat{R}(\omega,s,0)$ may be explicitely calculated from (5.28) and (8.17). Comparing (13.19) with the expression (13.4) for the precise reflection coefficient we conclude that the exponent in (13.19) presents the leading term of the corresponding exponent (13.4) as $\omega\to\infty$. Thus using the integral equation for the Ψ-function and the inversion formula

$$\bar{E} = 2\lim_{\omega\to\infty} \omega\Psi_{21}(\omega,x,t)\exp(-i\omega t)$$

it is possible to prove that the expression (13.14) approximates the precise solution E of system (13.1) with the order $O(x^{-\frac{1}{2}})$ for the residual term as $x\to\infty$.

Chapter 14. THE SCALING LIMIT IN TWO-DIMENSIONAL ISING MODEL

We consider here according to [5] the XY-model of Ising type on a finite lattice $M \times N$. The free energy in it is defined as usual

$$E(\sigma) = -E_1 \sum_{m=0}^{M-1} \sum_{n=0}^{N-1} \sigma_{mn} \sigma_{m+1,n} - \qquad (14.1)$$
$$- E_2 \sum_{m=0}^{M-1} \sum_{n=0}^{N-1} \sigma_{mn} \sigma_{m,n+1} ,$$

where $\sigma_{mn} = \pm 1$ - the value of spin variable in the (m,n)-point of the lattice.

The statistical sum Z_{MN} is given by the formula

$$Z_{MN} = \sum_{(\sigma)} exp(-\beta E(\sigma)) , \qquad (14.2)$$

where $\beta = (kT)^{-1} > 0$, k is the Bolzman constant, T is temperature. The sum (14.2) is calculated over all possible configurations of spins.

The two-point correlation function

$$\rho_2 = \langle \sigma_{m_1 n_1} , \sigma_{m_2 n_2} \rangle$$

is also defined by a standard manner

$$\rho_2 = \frac{1}{Z_{MN}} \sum_{(\sigma)} \sigma_{m_1 n_1} \sigma_{m_2 n_2} exp(-\beta E(\sigma)) . \qquad (14.3)$$

We label by T_c the temperature of the phase transition and by

ε, θ the new scaled temperatures

$$\varepsilon = |T - T_c|, \quad \theta = \frac{1}{2}\varepsilon\sqrt{M^2 + N^2}.$$

In many applications it is necessary to study the behaviour of the correlation function ρ_2 as $\varepsilon \to 0$, $M^2+N^2 \to \infty$. According to ref. [5] there exist the finite limits of ρ_2 while $T \to T_c$:

$$\rho_2^{\pm}(\theta) = \lim_{T \to T_c} \varepsilon^{-\frac{1}{4}} \langle \sigma_{00}, \sigma_{MN} \rangle .$$

Moreover, it is shown further that $\rho_2^{\pm}$ are expressed through the Painlevé functions of the III type

$$\rho_2^{\pm}(\theta) = C \left\{ \begin{matrix} sh \frac{\Phi}{2} \\ ch \frac{\Phi}{2} \end{matrix} \right\} exp \left\{ \frac{1}{4} \int \left[\Phi'^2(t) - sh^2 \Phi(t) \right] t \, dt \right\}, \tag{14.4}$$

where for Φ the PIII equation holds

$$\frac{d^2\Phi}{d\theta^2} + \frac{1}{\theta}\frac{d\Phi}{d\theta} = 2 sh\, 2\Phi . \tag{14.5}$$

The problem consists of studying the connection between $\rho_2^{\pm}$ in the two limit cases: $\theta \to 0$ and $\theta \to \infty$ (the scaling limit). By the formulae (14.4), (14.5) this problem is clearly reduced to the connection formulae produced in Chapter 8 for PIII equation. Putting $\Phi = -\ln \eta$ we extract the non-singular as $\theta \to \infty$ solution of eq. (14.5) which is of physical interest:

$$\eta(\theta) \to \begin{cases} B\theta^{\sigma}, & \theta \to 0, \\ 1 - \gamma\sqrt{2\pi}\,\frac{e^{-2\theta}}{\sqrt{\theta}}, & \theta \to \infty, \end{cases} \tag{14.6}$$

where B, σ, γ are real-valued parameters. The change of variables

$$\theta = -\frac{ix}{2}, \quad \Phi = \frac{iu}{2}$$

reduces the problem (14.5), (14.6) to the following one

$$\frac{d^2u}{dx^2} + \frac{1}{x}\frac{du}{dx} + \sin u = 0 ,$$

$$u(x) \to \begin{cases} 2i\sigma \ln x + 2i \ln B + \pi\sigma , & x \to i0 \\ \gamma\sqrt{8\pi}\, x^{-\frac{1}{2}} \exp i(x - \frac{\pi}{4}), & x \to i\infty . \end{cases} \tag{14.7}$$

One can show (we allow ourselves to omit computation) that the solution to the eq. PIII has analytical continuation to the sector $0 \leq \arg x \leq \leq \pi/2$, where the asymptotics (14.7) still remain valid. As a final reduction of the initial problem (14.6) we then obtain the following one

$$\frac{d^2u}{dx^2} + \frac{1}{x}\frac{du}{dx} + \sin u = 0$$

$$u(x) \to \begin{cases} 2i\sigma \ln x + 2i \ln B + \pi\sigma , & x \to 0 \\ \gamma\sqrt{8\pi}\, x^{-\frac{1}{2}} \exp i(x - \frac{\pi}{4}), & x \to \infty . \end{cases} \tag{14.8}$$

The relation between the asymptotics (14.8) and that of (8.7) is immediately established by putting

$$\alpha = 2\pi\gamma , \quad \varphi = -\frac{\pi}{2} , \quad \tau = 2i\sigma , \quad s = 2i \ln B + \pi\sigma .$$

The connection formulae (8.9) are now read

$$\sigma = \frac{2}{\pi} \arcsin \pi\gamma , \qquad 0 \leq \gamma < \frac{1}{\pi} ,$$

$$B = 2^{-3\sigma} \frac{\Gamma\left(\frac{1}{2} - \frac{\sigma}{2}\right)}{\Gamma\left(\frac{1}{2} + \frac{\sigma}{2}\right)} , \tag{14.9}$$

which coinside with the result of ref [5] .

It is worth mentioning that in ref. [5] the connection formulae (14.9) were obtained without any use of isomonodromic deformations. Instead of it the linear integral equation was used

$$\int_{-1}^{1} K_0(\theta|x-x'|) \begin{Bmatrix} f(x') \\ g(x') \end{Bmatrix} dx' = \begin{Bmatrix} ch\,\theta x \\ sh\,\theta x \end{Bmatrix}, \; -1<x<1 .$$

The Painlevé function $\Phi(\theta)$ was expressed through the solutions f, g and the connection formulae (14.9) was obtained by rather cumbersome iteration of multiple asymptotic series for f and g .

Chapter 15. QUASICLASSICAL MODE OF THE THREE-DIMENSIONAL WAVE COLLAPSE

The nonlinear Schrödinger equation

$$iU_t + \frac{1}{2}\Delta U + |U|^2 U = 0 \tag{15.1}$$

seems to be one of the most popular models in the physics of non-linear waves. It describes in particular the propagation of the spectrally narrow wave pack through a media with positive dispersion. The properties of the solutions depend essentially upon the dimension of the x-space $\mathbb{R}^d$. Since $d \geqslant 2$ the equation (15.1) describes a fundamental phenomena of the wave collaps, which make the amplitude $|U|$ to become infinite at some points (see [59], [60]). The wave collaps occurs during a finite interval of time and from the physical point of view it is a spontaneous concentration of the energy in a small domains of space, where it is subsequently dissipated. In some cases, such as the Langmure wave propagation in a plasma, the wave collaps might become the major process responsible for the wave energy dissipation (see [60]).

Consider the dynamics of collaps for a solution of the equation (15.1) in the case of $d=3$, which presents a certain interest for the plasma physics.

The equation (15.1) has two conservation laws

$$E = \int_{\mathbb{R}^3} |U|^2 dx , \tag{15.2}$$

$$H = \int_{\mathbb{R}^3} (|\nabla U|^2 - |U|^4) dx , \tag{15.3}$$

the values of which are determined by the initial data

$$U(x,0)=U_0(x), \quad U_0\in S(\mathbb{R}^3), \tag{15.4}$$

where $S(\mathbb{R}^3)$ is the Schwarz's space of functions. It is known (see [60]) that a sufficient condition for the development of collaps is a negative value of the Hamiltonian (15.3) $H<0$.

Unlike the one-dimensional case the three-dimensional Schrödinger equation (15.1) is not an exactly solvable one and it does not have the higher conservation laws. Nevertheless there exist the self-similar solutions, which are described by the Painlevé II equation ([20]). One of these solutions would be used below as a boundary-layer function matching two quasiclassical solutions describing the wave collaps inside and outside of a certain transitional domain. The asymptotics of the pure-imaginary solutions of PII equation, which have been found in Chapter 9, would play a significant role in the process of asymptotic matching.

Suppose the collaps to occur at the origin $x=0$ at the moment of time $t=t_0>0$. For simplicity consider a spherically symmetric case

$$U=U(r,t), \quad r=\sqrt{x_1^2+x_2^2+x_3^2}, \tag{15.5}$$

since a general solution U seems to transform into the spherically symmetric one near the point of collaps as it was confirmed by the results of [60] , [19] . Our aim is to construct an asymptotic expansion of the solution (15.5) uniform in x as $t\to t_0$.

Introduce according to [19] the self-similar variable

$$\xi=\frac{r}{l\tau^{2/5}}, \quad \tau=t_0-t, \quad \tau>0,$$

where l is a some positive constant, and consider the following asymptotic ansatz

$$U=U_{in}=\exp\{i\tau^{-1/5}\Phi(\xi)\}\cdot\left[\tau^{-3/5}f(\xi)+\tau^{-1/2}\alpha_1(\xi)e^{i\theta_1(\xi)}+\tau^{-1/2}\alpha_2(\xi)e^{-i\theta_2(\xi)}+\right.$$

$$\left.+\tau^{-2/5}\tilde{f}(\xi)+\tau^{-2/5}\tilde{\alpha}_1(\xi)e^{2i\theta_1(\xi)}+\tau^{-2/5}\tilde{\alpha}_2(\xi)e^{-2i\theta_2(\xi)}+\dots\right] \tag{15.6}$$

$$\theta_j(\xi)=\tau^{-1/5}\varphi(\xi)+\gamma_j\ln\tau+\beta_j(\xi)\ ,\quad j=1,2,$$

$$\tau\to 0 .$$

Substituting it into the spherically symmetric equation (15.1)

$$-iU_\tau+\frac{1}{2}U_{rr}+\frac{1}{r}U_r+|U|^2U=0\ , \tag{15.1}$$

we equate the terms of senior order in τ . We have

$$\tau^{-9/5}:\frac{1}{5}\Phi+\frac{2}{5}\xi\Phi_\xi+\frac{1}{2\ell^2}\Phi_\xi^2=f^2 \tag{15.7}$$

$$\tau^{-8/5}:\frac{3}{5}f+\frac{2}{5}\xi f_\xi+\frac{1}{\ell^2}\left(\Phi_\xi f_\xi+\frac{1}{2}\Phi_{\xi\xi}f+\frac{1}{\xi}\Phi_\xi f\right)=0 \tag{15.8}$$

$$\mp\tilde{\alpha}_{1,2}\varphi-2\cdot\ell^{-2}\tilde{\alpha}_{1,2}\varphi_\xi^2+f^2(\tilde{\alpha}_2+\tilde{\alpha}_1)+f(2\alpha_1\alpha_2+\alpha_{1,2}^2)=0 \tag{15.9}$$

$$f\tilde{f}=-\alpha_1^2-\alpha_2^2-\alpha_1\alpha_2 .$$

There are explicit solutions of the equations (15.7), (15.8), found in the work [19] :

$$\Phi(\xi)=\frac{3}{5}\ell^2\xi_c^2-\frac{1}{5}\ell^2\xi^2\ , \tag{15.9}$$

$$f(\xi)=\frac{\ell}{5}\sqrt{3(\xi_c^2-\xi^2)}\ ,\quad \xi<\xi_c\ , \tag{15.10}$$

where ξ_c is a certain positive constant.

The solution (15.9), (15.10) was called quasiclassical by the authors of [19] since the equation (15.8) might be considered as the Newton's equation describing the fall of a classical particle upon the center $\xi = 0$.

Equating the terms of order $\tau^{-\frac{17}{10}}$ and $\tau^{-\frac{3}{2}}$ in (15.1) we obtain the equations on the remainder parameters of (15.6)

$$\left.\begin{aligned} &\alpha_1\left(\frac{1}{5}\varphi + \frac{1}{2\ell^2}\varphi_\xi^2\right) = (\alpha_1 + \alpha_2) f^2 \\ &\alpha_2\left(-\frac{1}{5}\varphi - \frac{1}{2\ell^2}\varphi_\xi^2\right) = (\alpha_1 + \alpha_2) f^2 \end{aligned}\right\} , \qquad (15.11)$$

$$\left.\begin{aligned} &\varphi_\xi \alpha_{1\xi} + \left(\frac{1}{2}\varphi_{\xi\xi} + \frac{1}{\xi}\varphi_\xi - \frac{\ell^2}{10}\right)\alpha_1 = 0 , \\ &\varphi_\xi \alpha_{2\xi} + \left(\frac{1}{2}\varphi_{\xi\xi} + \frac{1}{\xi}\varphi_\xi + \frac{\ell^2}{10}\right)\alpha_2 = 0 \end{aligned}\right\} , \qquad (15.12)$$

$$\left\{\begin{aligned} &\varphi_\xi \beta_{1\xi} = \ell^2(\alpha_1^2 + 2\alpha_2^2 + \gamma_1) + \ell^2 f\tilde{f}\,\frac{4\alpha_1 + 2\alpha_2}{\alpha_1} + 2\ell^2 f(\alpha_2\tilde{\alpha}_2 + \alpha_1\tilde{\alpha}_1 + \alpha_2\tilde{\alpha}_1)/\alpha_1 \\ &\varphi_\xi \beta_{2\xi} = \ell^2(\alpha_2^2 + 2\alpha_1^2 + \gamma_2) + \ell^2 f\tilde{f}(4\alpha_2 + 2\alpha_1)/\alpha_2 + 2\ell^2 f(\alpha_1\tilde{\alpha}_1 + \alpha_2\tilde{\alpha}_2 + \alpha_1\tilde{\alpha}_2)/\alpha_2 \end{aligned}\right. \qquad (15.13)$$

The solution of the equations (15.12) with respect to α_1 and α_2 yields

$$\alpha_j(\xi) = \frac{C_j}{\xi\sqrt{\varphi_\xi}} \exp\left\{\pm \ell^2 \int_{\xi_c}^{\xi} (10\,\varphi_\xi)^{-1} d\xi\right\} , \qquad (15.14)$$

$$j = 1, 2 , \qquad C_j = const .$$

The compatibility condition for two equations (15.11) on a single function φ provides $C_1 = C_2$ in (15.14), so that φ satisfies the equation

$$\frac{1}{5}\mathcal{Y}+\frac{1}{2\ell^2}\mathcal{Y}_\xi^2=\Big(1+\exp\Big\{-\ell^2\int_{\xi_c}^{\xi}(10\mathcal{Y}_\xi)^{-1}d\xi\Big\}\Big)f^2 , \tag{15.15}$$

where the function $f^2(\xi)$ is defined by (15.10).

It is easy to prove the existence of solution of the equation (15.15) with the initial condition

$$\mathcal{Y}(\xi_c) = 0 \tag{15.16}$$

in a small neighbourhood of $\xi=\xi_c$. In fact the asymptotics of $\mathcal{Y}$ has the form

$$\mathcal{Y}(\xi)=A\cdot(\xi_c-\xi)^{3/2}+\delta(\xi-\xi_c),\quad \xi\to\xi_c-0 . \tag{15.17}$$

The substitution of this ansatz into (15.15) yields

$$A=-\frac{4\ell^2}{5\sqrt{3}}\sqrt{2\xi_c}\ ,\quad \delta(\xi-\xi_c)=O(\xi_c-\xi)^2 . \tag{15.18}$$

Since the phase function $\mathcal{Y}$ is constructed via the solution of the Cauchy problem (15.15), (15.16), the asymptotic formulae for $\alpha_j(\xi)$, $\beta_j(\xi)$ are obtained immediately from (15.14), (15.13) and (15.8'):

$$\alpha_j(\xi)=\frac{C}{2}(\xi_c-\xi)^{-1/4}+O\big(\sqrt{\xi_c-\xi}\,\big) , \tag{15.19}$$

$$\beta_j(\xi)=\frac{5C^2\sqrt{6}}{4\sqrt{\xi_c}}\ln(\xi_c-\xi)+\beta_0+O(\sqrt{\xi_c-\xi}),\ j=1,2, \tag{15.20}$$

$$\xi\to\xi_c-0,\ C,\ \beta_0=\text{const} .$$

The asymptotic formulae (15.17) - (15.20) show that the approximation (15.6) for the solution U is applicable only in the domain $0<\xi<\xi_c$. Moreover, the amplitudes α_j and phases β_j become singular as $\xi\to\xi_c$. In order to expand the solution (15.6) beyond the sphere of radius $\xi=\xi_c$ it is necessary to construct a boundary layer solution in the neighbourhood of this sphere. This solution would be matched with that of (15.6) as $\xi\to\xi_c-0$, while its asymptotics as $\xi\to\xi_c+0$ would provide the boundary value conditions for the approximation of U as $\xi>\xi_c$.

Introduce a "stratched" variable ζ for the boundary layer solution in the neighbourhood of $\xi=\xi_c$,

$$\xi=\xi_c+\tau^{\frac{2}{15}}\zeta .$$

A new approximation for the solution of (15.1) has the form

$$U=U_{trans}=\tau^{-\frac{8}{15}}u(\zeta)\exp\{i\tau^{-\frac{1}{5}}\Phi(\xi)\}\ , \tag{15.21}$$

where the phase function Φ is determined in (15.9). Substituting the ansatz (15.21) into the equation (15.1) one obtains an identity for the leading order terms while for the terms of order $\tau^{-8/5}$ the equation occers

$$\frac{6}{25}\xi_c l^2\zeta u+\frac{1}{2l^2}u_{\zeta\zeta}+u^3=0\,. \tag{15.22}$$

Clearly one recognises the Painlevé II equation in (15.22), which differs from the canonical form of PII only in its coeficients. The simple scaling of variables

$$\zeta=-\left(\frac{12l^4}{25}\xi_c\right)^{-1/3}\eta\,,\quad u=i\left(\frac{12l}{25}\xi_c\right)^{1/3}v(\zeta)\ , \tag{15.23}$$

reduces (15.22) to the familiar form

$$v_{\eta\eta} = \eta v + 2v^3 .$$

We are now going to choose the solution of (15.22) satisfying the matching condition with (15.6) in a transitional domain

$$\xi_c - \xi = O(\tau^{\frac{2}{15}-\varepsilon}), \ -\eta = O(\tau^{-\varepsilon})$$

$$0 < \varepsilon < \frac{2}{15} , \quad \tau \to 0 .$$

The matching condition may be expressed as follows

$$\lim_{\substack{\xi \to \xi_c - 0 \\ \eta \to -\infty}} | U_{in} \ U_{trans} | = 0 , \tag{15.24}$$

where the solutions U_{in}, U_{trans} are given by the formulae (15.6) , (15.21).

Bringing together the expansions (15.17) - (15-20), we derive the asymptotics for U_{in} as $\xi \to \xi_c - 0$ in the form

$$U_{in} = \exp\{i\tau^{-\frac{1}{5}}\Phi(\xi)\}\Big[\tau^{-\frac{3}{5}}\frac{\ell\sqrt{6\xi_c}}{5}\sqrt{\xi_c - \xi} +$$

$$+\tau^{-\frac{1}{2}} c(\xi_c - \xi)^{-\frac{1}{4}} \cos\Big\{\frac{4\ell^2}{5}\sqrt{\frac{2}{3}\xi_c}\, \tau^{-\frac{1}{5}}(\xi_c - \xi)^{\frac{3}{2}} - \gamma \ln\tau - \tag{15.25}$$

$$-\frac{5\sqrt{6}}{4\sqrt{\xi_c}} c^2 \ln(\xi_c - \xi) - \beta_0\Big\}\Big] \ \Big[1 + O(\tau^{1/5}) + O(\sqrt{\xi_c - \xi})\Big] ,$$

$$\tau \to 0 , \quad \xi \to \xi_c - 0 ,$$

where we have put $\gamma_1=\gamma_2=\gamma$.

The asymptotics (15.25) being rewritten through the variable $\zeta=+(\xi-\xi_c)\tau^{-2/15}$ show that the leading order term of asymptotics for the Painlevé function $u(\zeta)$ as $\zeta\to-\infty$ coinsides with $const\cdot\sqrt{-\zeta}$. This fact distinquishes the pure imaginary class of solutions of PII equation $v_{\eta\eta}=\eta v+2v^3$ since, according to the results of Chapter 9, those solutions are regular for all $\eta\in\mathbb{R}$ and have an asymptotics which includes $i\sqrt{\eta}$ as a leading term at infinity. Proceeding to the function $u(\zeta)$ via the transformation (15.23) we conslude that there exists a smooth real-valued solution of the equation (15.22) with the following asymptotics established by the formulae (9.27), (9.30) of Chapter 9

$$u(\zeta)=\ell\sqrt{\tfrac{6}{25}(-\zeta)\xi_c}+b\cdot\sqrt{\tfrac{2}{5}}(3\xi_c)^{1/4}(-\zeta)^{-\frac{1}{4}}\cos\Big\{\frac{2\sqrt{2}}{3}\cdot\frac{2\ell^2}{5}\cdot$$
$$\cdot\sqrt{3\xi_c}\cdot(-\zeta)^{3/2}-b^2\sqrt{2}\,\ln(-\zeta)^{3/2}+\varphi_-\Big\}\,\big[1+o(1)\big],\ \zeta\to-\infty, \tag{15.26}$$

$$u(\zeta)=a\sqrt{\tfrac{2}{5}}(3\xi_c)^{1/4}\zeta^{-1/4}\cos\Big\{\frac{2}{3}\cdot\frac{2\ell^2}{5}\sqrt{3\xi_c}\,\zeta^{3/2}+\frac{a^2}{2}\ln\zeta^{3/2}+ \tag{15.27}$$
$$+\varphi_+\Big\}\,\big[1+o(1)\big]\ ,\ \zeta\to+\infty.$$

Moreover, the results of Chapter 9 yield the connection formulae for the asymptotic parameters of (15.26) and (15.27). For example, the amplitudes b, a and the phase φ_+ are linked together by the constraint

$$b^2=\frac{3a^2}{4\sqrt{2}}-\frac{1}{2\pi\sqrt{2}}\ln 2\Big(sh\,\frac{\pi a^2}{2}\Big)-\frac{1}{\pi\sqrt{2}}\ln\Big|2\sin\Big\{\frac{3}{2}a^2\ln 2+$$

$$+\frac{\pi}{4}-\arg\Gamma\left(\frac{ia^2}{2}\right)-\varphi_+ +\frac{a^2}{4}\ln\left(\frac{12\ell^4}{25}\xi_c\right)\Big\}\Big| . \qquad (15.28)$$

The matching condition (15.24) provides a direct connection between C, β_0 and γ in (15.25) and b, φ_- in (15.26). Comparing the amplitudes and phases of (15.26) with that of (15.25) being rewritten in the variable ζ, we have

$$C=b\sqrt{\tfrac{2}{5}}(3\xi_c)^{1/4}, \quad \gamma=-\frac{C^2}{\sqrt{6\xi_c}}, \quad \varphi_-=-\beta_0 . \qquad (15.29)$$

Therefore the solutions U_{in} and U_{trans} become completely matched and the Painlevé solution $u(\zeta)$ is determined by its parameters b, φ_-. Applying now the connection formulae (9.), (9.), one of which is presented by (15.28), we obtain the values of a and φ_+ entering the asymptotics (15.27) as $\zeta\to+\infty$. This asymptotics in its turn have to be matched with a solution $U=U_{out}$ outside the sphere $\xi=\xi_c$. We proceed now to a construction of this solution. Its asymptotic ansatz has the form

$$\begin{aligned}U=U_{out}=\exp\{i\tau^{-\frac{1}{5}}\Phi(\xi)\}\big[\tau^{-\frac{1}{2}}\alpha_1(\xi)\exp i\{\tau^{-\frac{1}{5}}\varphi_1(\xi)+\\ +\gamma_1\ln\tau+\beta_1(\xi)\}+\tau^{-\frac{1}{2}}\alpha_2(\xi)\exp i\{\tau^{-\frac{1}{5}}\varphi_2(\xi)+\gamma_2\ln\tau+\\ +\beta_2(\xi)\}\big]\left[1+O(\tau^{\frac{1}{5}})\right], \quad \tau\to 0 ,\end{aligned} \qquad (15.30)$$

where the phase function $\Phi(\xi)$ is determined in (15.9). Substituting the expression (15.30) into the equation (15.1) and equating the leading order terms in τ, we get the equations on the phase functions

$$\frac{1}{5}\varphi_j + \frac{1}{2\ell^2}\varphi_{j\xi}^2 = -f^2(\xi)\,, \quad j=1,2\,, \tag{15.31}$$

where the function $f(\xi)$ is given by the formula (15.10). Note that the phase function $\Phi(\xi)$ satisfies a similar equation (15.7). The terms of order $\tau^{-3/2}$ yield the following equations on α_j, β_j:

$$\varphi_{j\xi}\alpha_{j\xi} + \left(\frac{1}{2}\varphi_{j\xi\xi} + \frac{1}{\xi}\varphi_{j\xi} - \frac{\ell^2}{10}\right)\alpha_j = 0\,, \quad j=1,2\,, \tag{15.32}$$

$$\left.\begin{aligned} \varphi_{1\xi}\,\ell^{-2}\beta_{1\xi} &= \gamma_1 + \alpha_1^2 + 2\alpha_2^2\,, \\ \varphi_{2\xi}\,\ell^{-2}\beta_{2\xi} &= \gamma_2 + 2\alpha_1^2 + \alpha_2^2 \end{aligned}\right\}, \tag{15.33}$$

γ_1, $\gamma_2 = const.$

Construct first a solution of the equation (15.31). Unlike (15.7) we need here a non-polynomial solution since the leading order term in the phase of (15.27) has the form $(\xi-\xi_c)^{3/2}$. The equation (15.31) may be reduced via the transform

$$\eta = \ell\sqrt{\tfrac{2}{5}}\,\xi\,, \quad \varphi_j = -\frac{3}{5}\ell^2\xi_c^2 + \psi\,,$$

to the form

$$\psi_\eta^2 + \psi = \frac{3}{2}\eta^2\,,$$

where the right-hand side arises due to the equality (15.10) $f^2(\xi) = \frac{3}{25}\ell^2(\xi_c^2 - \xi^2)$.

The general solution of the latter equation may be obtained in a parametric form (see [61])

$$\psi = -\nu^2 + \frac{3}{2}\eta^2 ,$$

$$(\eta + \frac{2}{3}\nu)^3 (\eta - \nu)^2 = A , \quad A = const .$$

Hence the phase functions φ_j has the representation

$$\left. \begin{aligned} \varphi_j(\xi) &= \frac{3}{5}\ell^2(\xi^2 - \xi_c^2) - \frac{2}{5}\ell^2 \mu_j^2(\xi) , \\ (\xi + \frac{2}{3}\mu_j)^3(\xi - \mu_j)^2 &= A_j , \ j = 1, 2 \end{aligned} \right\} . \qquad (15.34)$$

Assume the boundary condition for φ_j to be zero in the point $\xi = \xi_c$

$$\varphi_j(\xi_c) = 0 .$$

It implies the solutions φ_j to have the following asymptotics in the neighbourhood of the boundary

$$\begin{aligned} \varphi_1(\xi) &= \frac{2}{5}\ell^2 \frac{2}{\sqrt{3}} \sqrt{\xi_c}\, (\xi - \xi_c)^{3/2} + O(\xi - \xi_c)^2 , \\ \varphi_2(\xi) &= -\frac{2}{5}\ell^2 \frac{2}{\sqrt{3}} \sqrt{\xi_c}\, (\xi - \xi_c)^{3/2} + O(\xi - \xi_c)^2 , \\ &\xi \to \xi_c . \end{aligned} \qquad (15.35)$$

The representation (15.34) together with asymptotics (15.35) completely determine the phase functions φ_j . Thus the solutions of equations (15.32), (15.33) are calculated in terms of φ_j

$$\alpha_j(\xi) = \frac{C_j^+}{\xi\sqrt{\varphi_{j\xi}}} \exp\int_{\xi_c}^{\xi} \ell^2 (10\,\varphi_{j\xi})^{-1} d\xi ,$$

$$\beta_j(\xi)=\ell^2\int_{\xi_c}^{\xi}\frac{\gamma_j+2\alpha_1^2+2\alpha_2^2-\alpha_j^2}{\varphi_{j\xi}}\,d\xi+\overset{+}{B}_j\,, \tag{15.36}$$

$$\overset{+}{B}_j\,,\ \overset{+}{C}_j=const\,,\quad j=1,2\,.$$

A straightforward calculation provides (with the suitable choice of constants $\overset{+}{C}_j$) the asymptotics of α_j, β_j from the formulae (15.36)

$$\alpha_j(\xi)=\frac{1}{2}\overset{+}{C}(\xi-\xi_c)^{-1/4}+O(\sqrt{\xi-\xi_c})\,,$$

$$\beta_j(\xi)=(-1)^{j-1}\frac{5(\overset{+}{C})^2\sqrt{3}}{8\sqrt{\xi_c}}\ln(\xi-\xi_c)+\overset{+}{\beta}_{0j}+O(\sqrt{\xi-\xi_c})\,, \tag{15.37}$$

$$j=1,2\,,\quad \xi\to\xi_c\,.$$

The determination of constants $\overset{+}{C}$, $\overset{+}{\beta}_{0j}$ and γ_j proceeds in a similar way as above. Applying the asymptotic expansions (15.35), (15.37), one derives from (15.30)

$$U_{out}=exp\{i\tau^{-\frac{1}{5}}\Phi(\xi)\}\Big[\tau^{-\frac{1}{2}}\overset{+}{C}(\xi-\xi_c)^{-\frac{1}{4}}\cos\Big\{\frac{2\ell^2}{5}\cdot\frac{2}{\sqrt{3}}\sqrt{\xi_c}\,\cdot$$

$$\cdot\,\tau^{-\frac{1}{5}}(\xi-\xi_c)^{3/2}+\overset{+}{\gamma}\ln\tau+\frac{5(\overset{+}{C})^2\sqrt{3}}{8\sqrt{\xi_c}}\ln(\xi-\xi_c)+\overset{+}{\beta}_0\Big\}\Big[1+$$

$$+O(\tau^{1/5})+O(\sqrt{\xi-\xi_c})\Big]\,,\quad \tau\to 0\,,\ \xi\to\xi_c\,, \tag{15.38}$$

where it is assumed $\gamma_1=-\gamma_2=\overset{+}{\gamma}$, $\beta_{01}=-\beta_{02}=\overset{+}{\beta}_0$. The matching condition for (15.30) and (15.21) approximations of U reads

$$\lim_{\substack{\tau\to+\infty \\ \xi\to\xi_c+0}} | U_{out} - U_{trans} | = 0 .$$

Comparing the asymptotics (15.38) and (15.27) we have the connection formulae for the constants C^+, β_0^+, and a, φ_+ :

$$C^+ = a\sqrt{\tfrac{2}{5}}(3\xi_c)^{1/4} , \ \gamma^+ = -\frac{a^2}{10} , \ \beta_0^+ = \varphi_+ . \tag{15.39}$$

This completes the construction of the asymptotic solution of the equation (15.1) describing the process of the wave collaps.

REMARK 15.1. The proposed above asymptotic solution U depends upon five arbitrary real-valued constants: ℓ, ξ_c in the formulae (15.9), (15.10) together with C, γ, β_0 in the formulae (15.19), (15.20), (15.25). Clearly these constants depend on the initial condition (15.4) $U(x,0)=U_0(x)$. Moreover they might be calculated via two conservation laws (15.2) and (15.3). As a matter of fact, substituting the composite approximation $U=U_{in}+U_{out}$ into the integral (15.2) and integrating with respect to ξ, we have

$$E = 4\pi\int_0^\infty r^2 |U(r)|^2 dr =$$

$$= 4\pi\ell^3\int_0^{\xi_c} \xi^2 f^2(\xi)d\xi = \pi\cdot\left(\tfrac{2}{5}\right)^3 \ell^5 \xi_c^5 . \tag{15.40}$$

Thus it provides the equation on two constants ℓ, ξ_c. The other four equations are given by the second conservation law. The substitution of functions U_{in}, U_{out} and U_{trans} into (15.3) yield the sequence of equalities on the terms of orders $\tau^{+\frac{6}{5}-\frac{k}{5}-\frac{j}{2}}$, $2k+5j<12$, where k and j are integers. All but four of them appear to be identities, whereas others provide algebraic equations, complrting the equation (15.40) for the five constants. Unfortunately these are not so simple as (15.40), so we omit the corresponding calculations. Thus a complete asymptotic solution to the Cauchy problem (15.1), (15.4) is presented near the moment of collaps $t=t_0$.

APPENDIX 1

ON ASYMPTOTICS OF REGULAR SOLUTIONS FOR A SPECIAL KIND OF PAINLEVÉ V EQUATION

by B.I.Suleimanov

INTRODUCTION. A computation of one particle density matrix of impenetrable bosons at zero temperature has become a subject of discussion in recent papers [7] , [64] . The problem is reduced in [7] to a study of solutions of the equation

$$\Phi'' = \left[(\Phi')^2 - 1\right] \operatorname{ctg} \Phi + \frac{1}{x}(1 - \Phi') , \qquad (A.1)$$

which may be reduced via the substitution $y = \exp(2i\Phi)$ to certain kind of classical Painlevé V equation. More precisely, it is necessary to calculate an asymptotics as $x \to \infty$ of one-parameter class of solutions to equation (A.1) defined at $x = 0$ by the expansion

$$\Phi(x,a) = x - ax^2 + O(x^3) , \quad x \to 0 . \qquad (A.2)$$

The authors of [7] have studied numerically this solution and have proposed the following asymptotics for the case of $a > 1/\pi$:

$$\Phi(x,a) = -x + \tau(a) \ln x + x_0(\tau) + o(1) , \quad x \to +\infty , \qquad (A.3)$$

where

$$\tau(a) = -\frac{1}{\pi} \ln(\pi a - 1) , \qquad (A.4)$$

and $x_0(\tau) - \pi/2$ appears to be an odd function of τ .

The aim of this work is an analytical study of asymptotics of $\Phi(x,a)$ for all real-valued a . In particular the expansion

(A.3) would be proved and an explicit formula for $x_0(\tau)$ would be obtained.

The main result of this work may be formulated in the following

THEOREM. There are three types of asymptotics as $x \to \infty$ for the solution $\Phi(x,a)$ of (A.1), (A.2), depending on the value of a:

A) For $a < \frac{1}{\pi}$ the solution $\Phi(x,a)$ increases monotonically as $x \to \infty$,

$$\Phi(x,a) = x + \tau(a) \ln x + x_0(\tau) + o(1), \tag{A.5}$$

where

$$\tau(a) = \frac{1}{\pi} \ln(1 - a\pi), \tag{A.6}$$

$$x_0(\tau) = -2 \arg \Gamma\left(\frac{i\tau}{2}\right) + \tau \ln 2 - \pi \operatorname{sign} \tau, \tag{A.7}$$

$$\tau \neq 0, \quad x_0(0) = 0.$$

B) For $a = \frac{1}{\pi}$ the solution $\Phi(x,a)$ increases monotonically to a finite limit and

$$\Phi\left(x, \frac{1}{\pi}\right) = \frac{\pi}{2} + o(1),$$

C) For $a > \frac{1}{\pi}$ the asymptotic behaviour of $\Phi(x,a)$ is described by the formulae (A.3), (A.4), where

$$x_0(\tau) = \frac{\pi}{2} + 2 \arg \Gamma\left(\frac{i\tau}{2} - \frac{1}{2}\right) + \tau \ln 2 + l\pi, \quad l \in \mathbb{Z}. \tag{A.8}$$

The work consists of four paragraphs. In § 1 we reduce the equation (A.1) to another kind of PV equation, namely (see [11], [65 - 66]

$$h_{\tau\tau} = h_\tau^2 \frac{3h-1}{2h(h-1)} - \frac{1}{\tau}\left[h_\tau + h + \frac{h^{-1}}{8\tau}(h-1)^2\right], \tau = -\frac{1}{8} x^2. \tag{A.9}$$

This equation admits the $U-V$ pair representation, being a compatibility condition for two linear systems of equations in λ and τ respectively. By studying the equation in λ we are able to use the isomonodromic deformation technique similar to those developed in Chapter 7 for the case of PIII equation. However, there is one basic difference lying in the structure of a connection matrix. Its diagonal elements are determined not through the l o c a l behaviour of h the point $\tau = 0$ or $\tau = \infty$, but depend upon some integrals of h over the half-line $\tau > 0$, i.e. upon the g l o b a l properties of h. This fact seems to be an obstacle for construction of the connection matrix via the solution of the Painlevé equation. Fortunately the non-diagonal elements are completely determined by the local asymptotics of h as $\tau \to 0$ and $\tau \to \infty$. This circumstance together with some qualitative analysis of the Painlevé equation (A.1) make it possible to overcome the difficulty.

The connection matrix is calculated in § 2 via the initial condition (A.2). The asymptotics of solution for the cases A), B) is studied in § 3. The last paragraph contains an investigation of the case C).

§ 1. ISOMONODROMIC TECHNIQUE FOR PV EQUATION

As we have mentioned above the exponential transformation $y = \exp(2i\Phi)$ reduces the equation (A.1) to the following PV equation

$$y'' = (y')^2 \frac{3y-1}{2y(y-1)} - \frac{1}{x} y' + \frac{2i}{x} y + 2y \frac{y+1}{y-1} . \qquad \text{(A.10)}$$

Putting $x = 2it$ and $w = \dfrac{\sqrt{y}+1}{\sqrt{y}-1}$ we transform (A.10) into the special kind of PIII equation:

$$w'' = \frac{(w')^2}{w} - \frac{1}{t} w' + \frac{1}{t}(w^2 - 1) + \frac{w^4 - 1}{w} , \qquad \text{(A.11)}$$

which in its turn may be reduced to the equation (A.9) under the change of variables

$$t=\sqrt{2\tau}\ ,\quad h(\tau)=1-\frac{2}{w'-w^2+1}$$

Finally the relation between h and Φ reads

$$h(\tau)=1+\frac{2\sin^2\frac{\Phi(x)}{2}}{\Phi'(x)-1}\ . \tag{A.12}$$

The equation (A.9) is a compatibility condition for the following two systems of linear equations ([67]):

$$\frac{\partial\Psi}{\partial\lambda}=\left\{i\tau\begin{pmatrix}1 & 0\\ 0 & -1\end{pmatrix}+\frac{1}{\lambda}\begin{pmatrix}1/4 & B_1\\ C_1, & -1/4\end{pmatrix}+\frac{1}{\lambda^2}\begin{pmatrix}A_2 & B_2\\ B_2, & -A_2\end{pmatrix}\right\}\Psi, \tag{A.13}$$

$$\frac{\partial\Psi}{\partial\tau}=\left\{\begin{pmatrix}P_0 & Q_0\\ R_0, & -P_0\end{pmatrix}+\lambda\begin{pmatrix}P_1 & 0\\ 0, & -P_1\end{pmatrix}\right\}\Psi, \tag{A.14}$$

where

$$A_2=-\frac{i}{8}\frac{h+1}{h-1},\quad B_2=-\frac{1}{4}\frac{\sqrt{h}}{h-1},\quad B_1+C_1=-\frac{i}{2\sqrt{h}},\ B_1-C_1=+\frac{ih'\tau}{\sqrt{h}(1-h)},$$

$$P_0=\frac{1}{8\tau}\left(1+\frac{1}{h}\right),\quad P_1=-i,\quad \tau Q_0=B_1\ ,\quad \tau R_0=C_1\ . \tag{A.15}$$

According to the ideology of isomonodromic deformation method the equation in λ (A.13) plays the basic role in our study of PV equation (A.9). It has two irregular points $\lambda=0$ and $\lambda=\infty$. There exists a canonical solution $\overset{\infty}{\Psi}$ with an asymptotics as $\lambda\to\infty$ of the form

$$\overset{\infty}{\Psi}(\lambda) = \overset{\infty}{E}(\tau)(I + O(\lambda^{-1}))\lambda^{\sigma_3/4} e^{-i\tau\lambda\sigma_3}, \quad \arg\lambda = 0 . \qquad \text{(A.16)}$$

Here and further on the remainder term is estimated by taking maximum of absolute values for all matrix elements.

The assumption that $h(\tau)$ is a solution of PV equation (A.9) yields the following equation on $\overset{\infty}{E}(\tau)$ in (A.16)

$$\frac{d\overset{\infty}{E}}{d\tau} = P_0(\tau)\sigma_3 \overset{\infty}{E} ,$$

where P_0 is defined in (A.15). Hence we have put

$$\overset{\infty}{E}(\tau) = \tau^{\sigma_3/8} \exp\left\{ \frac{\sigma_3}{8} \int_{-b}^{\tau} \frac{dt}{h(t)\cdot t} \right\} = \tau^{\sigma_3/8} \exp\{\sigma_3 \mathcal{J}(\tau)\} , \qquad \text{(A.17)}$$

where $b > 0$ is constant.

On the other hand, the system (A.13) has another canonical solution $\overset{\circ}{\Psi}$ in a neighbourhood of the origin ([48]):

$$\overset{\circ}{\Psi}(\lambda) = H(\tau)\overset{\circ}{E}(\tau)(I + O(\lambda))\cdot\lambda^{\sigma_3/4} e^{i\sigma_3/\lambda}, \quad \lambda \to 0 , \quad \arg\lambda = 0 . \qquad \text{(A.18)}$$

Here the coefficients H and $\overset{\circ}{E}$ have the form

$$H(\tau) = \frac{1}{\sqrt{h(\tau) - 1}} (i\sigma_3\sqrt{h} + \sigma_1) , \qquad \text{(A.19)}$$

$$\overset{\circ}{E}(\tau) = \tau^{\sigma_3/8} \exp\{-\sigma_3 \mathcal{J}(\tau)\} . \qquad \text{(A.20)}$$

The normalization matrices (A.19), (A.20) are chosen in such a way that $\overset{\circ}{\Psi}$ satisfies the system (A.14). Define the connection matrix in a standard way

$$\overset{\infty}{\Psi}(\lambda) = \overset{\circ}{\Psi}(\lambda) Q. \tag{A.21}$$

The isomonodromic condition $\frac{dQ}{d\tau} = 0$ holds here due to the fact, that both $\overset{\infty}{\Psi}$ and $\overset{\circ}{\Psi}$ are the solutions of (A.14) (see the proof of the theorem 3.1 in Chapter 3). According to the methods developed in Chapter 7, 8 of the main text, we use the invariance of Q in order to obtain the asymptotic parameters τ and x_0 in (A.3),(A.5) as the functions of initial condition a in (A.2).

§ 2. CALCULATION OF THE CONNECTION MATRIX AS A FUNCTION OF a

In this paragraph we assume τ to be a small positive parameter and Φ satisfying the initial condition (A.2). Hence via the transformation (A.12) we have

$$h = 1 - \frac{x}{4a} + O(x^2) = 1 - \frac{i}{c}\tau^{1/2} + O(\tau) \ , \quad c = a\sqrt{2}. \tag{A.22}$$

The existence of such a solution for the equation (A.9) was established in [68] via the construction of a power series converging near origin $\tau = 0$. Combining this result with the facts, that Φ is real-valued and $y = exp(2i\Phi)$ is meromorfic function, we prove the analiticity of Φ for all x .

We are now going to study the direct problem of the monodromy theory for the system (A.13), i.e. shall calculate the connection matrix (A.21) while putting $\tau \to 0$ and $h(\tau)$ satisfying (A.22). It is convenient to seek a solution Ψ in the form

$$\Psi(\lambda) = H(\tau) V(\eta) \ , \quad \eta = \lambda^{-1} \ , \tag{A.23}$$

(compare with (A.19)), so that the system (A.13) takes the form

$$\frac{dV}{d\eta}=\left\{\frac{8\tau}{\eta^2}\begin{pmatrix}-A_2 & B_2\\ B_2, & +A_2\end{pmatrix}-\frac{1}{\eta}\begin{pmatrix}1/4 & C_1\\ B_1, & -1/4\end{pmatrix}-\frac{i}{8}\sigma_3\right\}V, \tag{A.24}$$

where the coefficients are being substituted by their asymptotics according to (A.15) and (A.22):

$$A_2=\frac{c}{4\varepsilon}+O(1),\quad B_2=-\frac{ic}{4\varepsilon}+O(1),$$
$$B_1=-\frac{i}{2}+O(\varepsilon),\quad C_1=c\varepsilon+O(\varepsilon^2), \tag{A.25}$$
$$c=a\sqrt{2},\quad \tau=\varepsilon^2 .$$

We construct first an asymptotics of $\overset{\circ}{\Psi}(\lambda,\varepsilon)$ as $\varepsilon\to 0$, $\lambda\to\infty$.

PROPOSITION 1. Let $\frac{1}{2}\varepsilon<\eta<2\varepsilon$, then the following estimate holds

$$\overset{\circ}{\Psi}(\lambda,\varepsilon)=\varepsilon^{-1/2}\sqrt{ic}\,\eta^{1/4}\begin{pmatrix}\frac{\sqrt{\pi}}{2}\exp(-\mu+\frac{i\pi}{4}), & \exp(\mu)\\ -i\frac{\sqrt{\pi}}{2}\exp(-\mu+\frac{i\pi}{4}), & -i\exp(\mu)\end{pmatrix}(I+O(\varepsilon^{1/4})),\ \varepsilon\to 0, \tag{A.26}$$

where

$$\mu=\frac{1}{8}\int_{-b}^{0}\left(\frac{1}{h(\tau)}-1\right)\frac{d\tau}{\tau}+\ln b^{-\frac{1}{8}} . \tag{A.27}$$

PROOF. Discarding the term of order $O(\frac{\varepsilon^1}{\eta^2})$, consider an abridget system (A.24)

$$\frac{d\tilde V}{d\eta}=\left\{\frac{i}{8}\sigma_3-\frac{1}{\eta}\begin{pmatrix}1/4 & C_1\\ B_1 & -1/4\end{pmatrix}\right\}\tilde V . \tag{A.28}$$

Its solutions may be expressed in terms of the Whittaker functions $W(\kappa, m, \pm\frac{i\eta}{4})$ (see, for example, [53]):

$$\tilde{V}=\eta^{-1/2}\begin{pmatrix} d_1\cdot W(\frac{1}{4},m,-\frac{i}{4}\eta), & C_1\cdot d_2\cdot W(-\frac{1}{4},m,\frac{i}{4}\eta) \\ B_1\cdot d_1 W(-\frac{3}{4},m,-\frac{i}{4}\eta), & d_2 W(\frac{3}{4},m,\frac{i}{4}\eta) \end{pmatrix}, \tag{A.29}$$

where

$$m=\frac{1}{4}\sqrt{1+16B_1C_1}=\frac{1}{4}-ic\varepsilon+O(\varepsilon^2),\quad \varepsilon\to 0 . \tag{A.30}$$

The normalization constants d_1, d_2 in (A.29) are chosen according to the boundary value condition (A.20), taking into account (A.17), (A.23) and the asymptotics of $W(\kappa, m, \pm\frac{i}{4}\eta)$ as $\eta\to\infty$. As a result we have

$$d_1=(4i)^{1/4}\exp(-\mathcal{J}(\tau))\cdot\tau^{1/8}=(4i)^{1/4}\exp(-\mu)[1+O(\varepsilon)], \tag{A.31}$$

$$d_2=(-4i)^{3/4}\exp(\mathcal{J}(\tau))\cdot\tau^{-1/8}=(-4i)^{3/4}\exp\mu\,[1+O(\varepsilon)],$$

Let $\overset{\circ}{V}$ be the precise solution of (A.24), $\overset{\circ}{V}=H^{-1}(\tau)\overset{\circ}{\Psi}(\eta,\varepsilon)$. The construction of $\tilde{V}$ provides the estimate

$$\tilde{V}^{-1}\overset{\circ}{V}=I+O(\eta^{-1}),\quad \eta\to\infty,\quad \eta\in\mathbb{R}. \tag{A.32}$$

It is easy to prove an estimate

$$\tilde{V}(\eta,\varepsilon)=\overset{\circ}{V}(\eta,\varepsilon)(I+O(\varepsilon)),\quad |\eta|>const=O(1), \tag{A.33}$$

writing down an integral equation for the components of $\tilde{V}^{-1}\overset{\circ}{V}$ and applying the perturbation theory approximations for small ε and finite η. In order to obtain necessary approsimation of $\overset{\circ}{V}$ for small η, we need an information of asymptotic behaviour of $\tilde{V}$

as $\eta \to 0$. The well-known asymptotics of the Whittaker functions $W(k, m, \pm\frac{i}{4}\eta)$ (see [53]) together with formulae (A.29), (A.31) provide the following expansion

$$\tilde{V} = \begin{pmatrix} \eta^{-1/4} - c\varepsilon\sqrt{\pi i}\,\eta^{1/4} , & 2c\varepsilon(\sqrt{-\pi i}\,\eta^{-1/4} - \eta^{1/4}) \\ -i\eta^{-1/4} + \frac{1}{2}\sqrt{\pi i}\,\eta^{1/4} , & 2c\varepsilon i\sqrt{\pi i}\,\eta^{-1/4} + \eta^{1/4} \end{pmatrix} exp(-\mu\sigma_3)(I +$$

$$+ O(\varepsilon \ln|\eta|) + O(\eta)), \quad \varepsilon \to 0, \ \eta \to 0. \tag{A.34}$$

The integral equation for the function $\tilde{V}^{-1}\overset{\circ}{V}$, which is equivalent to the system (A.24), has the form

$$(\tilde{V}^{-1}\overset{\circ}{V})(\eta) = I + \varepsilon \int_{\infty}^{\eta} G(\xi, \varepsilon)(\tilde{V}^{-1}\overset{\circ}{V})(\xi) d\xi , \tag{A.35}$$

where

$$G(\eta, \varepsilon) = 8\tilde{V}^{-1}(\eta)\varepsilon \begin{pmatrix} A_2 , & B_2 \\ -B_2 , & -A_2 \end{pmatrix} \eta^{-2} \tilde{V}(\eta) ,$$

so that the formulae (A.25), (A.34) yield the asymptotics

$$G(\eta, \varepsilon) = -\frac{8i}{\eta^2}\tilde{V}^{-1}(\eta)\begin{pmatrix} i & 1 \\ 1 & -i \end{pmatrix}\tilde{V}(\eta) = O(\varepsilon\eta^{-5/2} + \eta^{-7/4}), \ \varepsilon \to 0, \ \eta \to 0 . \tag{A.36}$$

Evaluating now the solution of (A.35) via the perturbation theory method for small ε and η, one gets the estimate

$$\overset{\circ}{V} = \tilde{V} \cdot (I + O(\varepsilon^{1/4})), \ \frac{\varepsilon}{2} < \eta < 2\varepsilon, \ \varepsilon \longrightarrow 0 ,$$

which yields the statement of the Proposition 1.

Now we proceed to an investigation of asymptotic behaviour of $\overset{\infty}{\Psi}$ as $\varepsilon \to 0$. It is convenient here to rewrite the system (A.13) in a new variable

$$z = \varepsilon^2 \lambda .$$

Thus we have

$$\frac{d\Psi}{dz} = \left\{ -i\sigma_3 + \frac{1}{z} \begin{pmatrix} 1/4 & , & B_1 \\ C_1 & , & -1/4 \end{pmatrix} + \frac{\varepsilon^2}{z^2} \begin{pmatrix} A_2 & , & B_2 \\ B_2 & , & -A_2 \end{pmatrix} \right\} \Psi .$$

Discarding just as above the term of order $\mathcal{O}\left(\frac{\varepsilon^1}{z^2}\right)$, consider an abridged system

$$\frac{d\tilde{\Psi}}{dz} = \left\{ -i\sigma_3 + \frac{1}{z} \begin{pmatrix} 1/4 & , & B_1 \\ C_1 & , & -1/4 \end{pmatrix} \right\} \tilde{\Psi} ,$$

which in fact coinsides with that of (A.28). Its solutions again may be expressed via the Whittaker functions

$$\tilde{\Psi} = z^{-1/2} \begin{pmatrix} f_1 \cdot W(\frac{3}{4}, m, 2iz) , & -f_2 B_1 W(-\frac{3}{4}, m, -2iz) \\ -f_1 C_1 W(-\frac{1}{4}, m, 2iz), & f_2 W(\frac{1}{4}, m, -2iz) \end{pmatrix} .$$

The normalization constants f_1, f_2 are chosen here according to the boundary value condition (A.16) together with the constraint (A.17) and an asymptotics of the Whittaker functions as $z \to \infty$. As a result one obtains

$$f_1 = (2i)^{-3/4} \tau^{-1/8} \exp(\mathcal{J}(\tau)) = (2i)^{-3/4} \exp(\mu)(1 + \mathcal{O}(\varepsilon)) ,$$

$$f_2 = (2i)^{-1/4} \tau^{1/8} \exp(-\mathcal{J}(\tau)) = (2i)^{-1/4} \exp(-\mu)(1 + \mathcal{O}(\varepsilon)), \quad \varepsilon \to 0 .$$

Repeating the same reasoning as above one arrives to the following estimate

$$\overset{\infty}{\Psi} = \tilde{\Psi}(z, \varepsilon)\left(I + \mathcal{O}(\varepsilon^{1/4}) \right) = \left[\begin{pmatrix} z^{1/4} & iz^{-1/4} + i\sqrt{\frac{\pi i}{2}}\, z^{1/4} \\ 0 & z^{1/4} \end{pmatrix} + \right.$$

$$+ O(\varepsilon z^{-1/4}) \Big] \exp(\mu\sigma_3)\big[I + O(\varepsilon^{1/4})\big] , \ \frac{\varepsilon}{2} < z < 2\varepsilon .$$

This proves the following

PROPOSITION 2. Let $\frac{\varepsilon}{2} < z < 2\varepsilon$, then the estimate holds

$$\overset{\infty}{\Psi}(z,\varepsilon) = \begin{pmatrix} 0 & i z^{-1/4} \exp(\mu) \\ 0 & z^{-1/4} \exp(-\mu) \end{pmatrix} \big[I + O(\varepsilon^{1/4})\big], \ \varepsilon \to 0 . \tag{A.37}$$

We are able now to calculate the connection matrix Q . Returning to the variable $\lambda = \eta^{-1} = \varepsilon^{-2} z$, one concludes that the domains $\frac{\varepsilon}{2} < z < 2\varepsilon$ and $\frac{\varepsilon}{2} < \eta < 2\varepsilon$ coinside, so the representations (A.26) and (A.37) may be substituted instead of $\overset{\circ}{\Psi}$, $\overset{\infty}{\Psi}$ into the constraint

$$\overset{\circ}{\Psi} = \overset{\infty}{\Psi} Q^{-1} .$$

As a result are obtains explicit expressions for the matrix elements of Q :

$$Q_{11} = -i\tau^{1/4} \exp\Big\{ \frac{1}{4} \int_{-b}^{\tau} (h^{-1}(\tau))\tau^{-1} d\tau \Big\}, \tau \to 0 ,$$

$$Q_{21} = i 2^{-3/4} \sqrt{a\pi} . \tag{A.38}$$

The first formula (A.38) shows that the diagonal element Q_{11} depends nonlocally upon the potential $h(\tau)$. The reason lies in our choice of normalization conditions (A.16), (A.18), where the diagonal of Ψ-function is determined via the integral $\mathcal{J}(\tau)$ of $\tau^{-1} h^{-1}(\tau)$ taken along the interval $[\tau, -b]$. Since the integration constant b is fixed in the formula (A.17), it implies the dependence of Q upon b .

Note that in § 3 the element Q_{11} would be calculated in terms of integral $\mathcal{J}(-\infty)$, taken along the interval $[-b,-\infty]$. Thus the invariance of Q in τ makes it possible to calculate the explicit value of integral $\mathcal{J}$ along the half-line $[0,-\infty)$.

§ 3. ASYMPTOTICS OF THE INCREASING SOLUTIONS AS $x\to\infty$

In this paragraph we return for a while from (A.9) to the initial equation (A.1) and discuss the existence of its solutions with the asymptotics, increasing at infinity.

LEMMA 1. The solution $\Phi(x,a)$ of the initial value problem (A.1), (A.2) is a continuous function of a together with all its derivatives.

PROOF. The statement is true in a neighbourhood of the origin because $\Phi(x,a)$ is there an analytic function. For all the rest values of x it follows from the results obtained in [68], where the Painlevé equation (A.10) has been reduced to a Hamiltonian system

$$\begin{cases} \dfrac{dy}{dx}=\dfrac{\partial H}{\partial z} \\[2ex] \dfrac{dz}{dx}=-\dfrac{\partial H}{\partial y} \end{cases}$$

with a polynomial Hamiltonian $H=2iyz+\frac{1}{x}y(y-1)^2z^2$. The lemma is proved.

LEMMA 2. If $a<0$, then every solution of (A.1), (A.2) is monotonically increasing function, and $\Phi_x(x,a)\geqslant 1$. The equality $\Phi_x(x,a)=1$ holds in such point x_n, where $\Phi(x_n,a)=\pi n,\ n\in\mathbb{Z}$.

PROOF. The assumption that $\Phi(x^*,a)\neq\pi n$ and simultaneously $\Phi_x'(x^*,a)=1$ contradicts with the existence of solution having the form $\Phi(x)=x+\Phi(x^*,a)-x^*$. At the same time the point x_*, where $\Phi_x'(x_*,a)=1$ and $\Phi(x_*,a)=\pi n$ appears to be the bending point as $\Phi''(x_*, a)=0$ and $\Phi'''(x_*, a)\neq 0$. The lemma is proved.

A quite similar proof holds for the following

LEMMA 3. If $a>0$, then every solution of (A.1), (A.2) satisfies the estimate $\Phi'_x(x, a) \leqslant 1$, and the equality here is reached only at such points x_ℓ , where $\Phi(x_\ell, a)=\pi\ell$, $\ell\in\mathbb{Z}$.

Further properties of increasing solutions are established in forthcoming lemmas 4 - 7.

LEMMA 4. Let $\Phi(x)$ be a solution of the equation (A.1) defined on the interval $[x_1, x_0]$ and posessing the following properties:

$$\Phi(x_0)=\pi\ell,\quad \ell\in\mathbb{Z},$$

$$0<\Phi'(x)\leqslant 1,\quad \Phi'(x)\not\equiv 1,\quad \Phi(x_1)>(\ell-1)\pi .$$

Assume $\tilde{\Phi}(x)$ to be a smooth solution of equation

$$\tilde{\Phi}''=[(\tilde{\Phi}')^2-1]\,\mathrm{ctg}\,\tilde{\Phi}, \tag{A.39}$$

defined on the same interval and satisfying the condition

$$0<\tilde{\Phi}'(x)\leqslant 1,\quad \tilde{\Phi}(x_0)=\pi\ell,\quad \tilde{\Phi}(x_0)<\Phi(x) \quad \text{for } x\to x_0-0.$$

Then $\tilde{\Phi}(x)<\Phi(x)$ on the whole semi-interval $[x_1, x_0)$.

PROOF. Let x_2 be the rightest point where $\Phi(x_2)=\tilde{\Phi}(x_2)$, $\Phi'(x_2)\leqslant\tilde{\Phi}'(x_2)<1$. The latter estimate follows from the uniqueness theorem for a solution of the Cauchy problem and the fact that $\tilde{\Phi}'(x_0)=$ $=\Phi(x_0)=1$. Consider the inverse functions $z(\Phi)=\Phi^{-1}$ and $\tilde{z}(\Phi)=\tilde{\Phi}^{-1}$ on an interval $[\ell\pi, \alpha]$, where $\alpha=\Phi(x_2)$. They satisfy the equations

$$z''=[(z')^3-z']\,\mathrm{ctg}\,\Phi+\frac{1}{z}[(z')^2-(z')^3],$$

$$\tilde{z}'' = [(\tilde{z}')^3 - \tilde{z}']\, ctg\, \Phi .$$

Under the conditions of the lemma we have

$$z'' < [(z')^3 - z']\, ctg\, \Phi \quad \text{as} \quad l\pi < \Phi \leq \alpha .$$

Taking the function $u = \tilde{z} - z$ one derives the inequality

$$u''_{\Phi\Phi} + u'_{\Phi}\, ctg\, \Phi [1 + (z'_{\Phi} + \tilde{z}'_{\Phi}) + z'_{\Phi}\tilde{z}'_{\Phi}] > 0 ,$$

which contradicts with the existence of such Φ_0 that $u'_{\Phi}(\Phi_0) = 0$ and $u''_{\Phi\Phi}(\Phi_0) \leq 0$.

The lemma is proved.

The similar reasoning proves the next lemma 5.

LEMMA 5. Let $\Phi(x)$ be a solution of the equation (A.1) defined on an interval $[x_0, x_1]$. Suppose the following conditions to be satisfied

$$\Phi(x_0) = \pi l , \quad l \in \mathbb{Z} ,$$

$$0 < \Phi'(x) \leq 1 , \quad \Phi'(x) \not\equiv 1 , \quad \Phi(x_1) < \pi(l-1) .$$

Let $\tilde{\Phi}(x)$ to be a solution of the equation (A.39) with the similar properties

$$\tilde{\Phi}(x_0) = \pi l, \quad 0 < \tilde{\Phi}'(x) \leq 1, \quad \tilde{\Phi}(x) < \Phi(x) \quad \text{for } x \to x_0 + 0 .$$

Then the inequality $\tilde{\Phi}(x) < \Phi(x)$ holds for all x of the semi-interval $(x_0, x_1]$.

LEMMA 6. Let $\Phi(x, a)$ to be a smooth solution of the problem (A.1), (A.2), satisfying the following conditions on an interval $[x_0, x_1]$:

$$\Phi'(x) > 0 , \quad \Phi'(x) \not\equiv 1 , \quad \Phi(x_0) = \pi(l-1), \quad \Phi(x_1) = \pi l , \quad l \in \mathbb{Z} .$$

Then there exists $x^* > x_1$ such that the following estimates hold

$$\Phi'(x) > 0 \ , \quad x > x_1 \ ,$$

$$|\Phi'(x) - 1 + \frac{8\kappa^2}{x} \operatorname{sign} a \sin^2 \Phi(x)| \leq \frac{C(\kappa^2, a)}{x} , \quad x > x^* ,$$

where $C(\kappa^2, a)$ and x^* are continuous function of κ^2, a .

PROOF. Consider first the case $a > 0$. In the neighbourhood of $x = x_1$ due to analyticity of Φ we have

$$\Phi(x) = \pi l + x - x_1 - d(x - x_1)^3 + O(x - x_1)^4) ,$$
$$\Phi'(x) = 1 - 3d(x - x_1)^2 + O(|x - x_1|^3) , \tag{A.40}$$

where $0 < d < \frac{1}{6}$, which is provided by Lemma 4 while comparing the expansions (A.40) with those for the function $\tilde{\Phi}(x)$, satisfying the equation $(\tilde{\Phi}')^2 = \cos^2 \tilde{\Phi}$ and the boundary value conditions $\tilde{\Phi}(x_1) = \pi l$, $\tilde{\Phi}'(x_1) = 1$. The results of lemmas 4 and 5 together with the symmetry

$$\tilde{\Phi}(x_1 + \xi) - l\pi = l\pi - \tilde{\Phi}(x_1 - \xi)$$

for the function $\tilde{\Phi}(x)$, satisfying the equation $\tilde{\Phi}' = \sqrt{1 - 6d \sin^2 \tilde{\Phi}}$ and conditions $\tilde{\Phi}(x_1) = \pi l$, $\tilde{\Phi}'(x_1) = 1$, yield the existence of a point $x_2 > x_1$ such that

$$\Phi(x_2) = \pi(l + 1)$$

and $\Phi'(x) > 0$ for all $x > x_0$.

Proceeding to the proof of the second estimate, consider the function

$$u(x) = x(1 - \Phi'(x)) .$$

It satisfies the inequality $0 \leq u(x) < x$ and the asymptotics

$$u(x) = 3d x_1 (x - x_1)^2 + O(x - x_1)^3 , \tag{A.41}$$

following from (A.40).

Substituting $\Phi'(x) = 1 - \frac{u(x)}{x}$ into the equation (A.1) and inverting the function Φ ($\Phi'(x)>0$) , we have

$$\frac{du}{d\Phi} = 2u \operatorname{ctg} \Phi + \frac{1}{x} u u'_{\Phi} - \frac{1}{x} u^2 \operatorname{ctg} \Phi . \tag{A.42}$$

It is equivalent to the integral equation

$$u(x) = 3dx_0 \sin^2\Phi + \frac{u^2}{2x} + \sin^2\Phi \int_{x_0}^{x} \frac{u^2(y)}{2y^2 \sin^2\Phi(y)} dy , \tag{A.43}$$

which provides an estimate

$$f(x) \leq 6dx_0 + \int_{x_0}^{x} \frac{f^2(y)}{y^2} dy ,$$

where $f(x) = u(x) \cdot |\sin\Phi(x)|^{-1}$.

Applying the Bihari's lemma (see [69]) to this inequality, one obtains

$$f(x) \leq \frac{6dx_1}{1-6dx_1(x_1^{-1}-x^{-1})} < \frac{x_1}{1-6d} ,$$

as $x > x_1$ and $6d < 1$.

Thus the integral equation (A.43) may be presented in the form

$$u(x) = 8\kappa^2 \sin\Phi(x) + \frac{u^2(x)}{2x} - \sin^2\Phi(x) \int_{x}^{\infty} \frac{u^2(y)}{2y^2 \sin^2\Phi(y)} dy ,$$

where $8\kappa^2 = 6dx_0 + \int_{x_0}^{\infty} \frac{f^2(y)}{2y^2} dy$, so that $f(x) \leq 16\kappa^2$, which proves the lemma for the case of $a>0$. For the remainder case $a<0$ the second estimate of the lemma is proved in a similar way. The integral equation (A.43) here takes the form

$$u(x) = 2a \sin^2\Phi(x) + \frac{u^2(x)}{2x} + \sin^2\Phi(x) \int_{0}^{x} \frac{u^2(y)\, dy}{2y^2 \sin^2\Phi(y)} ,$$

It yields the estimate $|u(x)\sin^{-1}\Phi(x,a)| \leq 2|a|$ which completes the proof.

Introduce now the set L consisting of all real-valued a such that $\Phi(x,a)$ increases monotonically as $x>0$ and intersects the margin $\Phi=\pi$. Obviously this set contains all negative a.

LEMMA 7. The set L is open.

PROOF. Let $x^*>0$ be the point where $\Phi(x,a)$ has its first maximum. The right-hand side of the equation (A.1) shows that $\Phi(x^*,a)<$ $<\frac{\pi}{2}$, and thus the statement of lemma follows from the lemmas 1 and 6.

We return to the study of the connection matrix for the equation (A.9), since we have established the existence of solutions increasing as $x\to\infty$ for the equation (A.1). The asymptotics (A.5) via the transformation (A.12) would be substituted into the system in λ (A.13) in order to calculate the connection matrix as $x\to\infty$. Comparing it with those calculated for $x=0$ we shall obtain the connection formulae stated in the Theorem above.

For convenience we unite the studies of the two cases, $a>0$ and $a\leq 0$, putting

$$\Phi'(x,a)=1+\frac{8k^2(a)}{x}\sin^2\Phi(x,a)+O(x^{-2}),\quad x\to\infty, \tag{A.44}$$

where $k=k(a)$ is real-valued for $a\leq 0$ and pure imaginary for $a>0$. The representation (A.44) was established in Lemma 6 for $a\in L$.

Introduce now a small parameter $\varepsilon^2=x^{-1}$ and expand the matrix elements of (A.13) in ε with respect to the relations (A.12), (A.44). We have

$$A_2=-\frac{i}{8}-ik^2\varepsilon^2\cos^2 S+O(\varepsilon^4),$$

$$B_2 = -\varepsilon \frac{K \cos S}{2} + O(\varepsilon^3),$$

$$B_1 = -\varepsilon^{-1} \frac{iK \sin S}{2} + O(\varepsilon), \tag{A.45}$$

$$C_1 = \varepsilon^{-1} \frac{iK \sin S}{2} + O(\varepsilon), \qquad \varepsilon \to 0,$$

where $S = \Phi/2$.

Substituting (A.45) into the system (A.13) and rescaling the variable λ,

$$\lambda = \xi \varepsilon^2 ,$$

we present the system (A.13) in the form

$$\varepsilon^2 \frac{d\Psi}{d\xi} = \Big[\frac{i}{8}\Big(1 - \frac{1}{\xi^2}\Big)\sigma_3 - \varepsilon \frac{K}{\xi^2}(\sigma_1 \cos S - \xi \sigma_2 \sin S) +$$

$$+ \varepsilon^2 \Big(\frac{1}{4\xi} - \frac{K^2 \cos^2 S}{\xi^2}\Big)\sigma_3 + \varepsilon^3 R(\xi, \varepsilon)\Big]\Psi , \tag{A.46}$$

where the remainder term has an estimate

$$R = \big(O(\varepsilon \xi^{-2})\sigma_3 + O(\xi^{-1})\sigma_1\big), \quad \varepsilon \to 0, \quad |\xi| > 1 . \tag{A.47}$$

The asymptotic solution of the system (A.46) as $\varepsilon \to 0$ may be constructed via the WKB-approximation technique. The WKB-ansatz here has the form

$$\hat{\Psi}(\xi, \varepsilon) = W_0(\xi) exp[i\theta\sigma_3] + \varepsilon W_1(\xi) exp[-i\theta\sigma_3] , \tag{A.48}$$

where

$$\theta = \frac{1}{8\varepsilon^2}\Big(\xi + \frac{1}{\xi}\Big) .$$

Substituting (A.48) into (A.46) and equating the terms of orders ε

and ε^2 one obtains the equation on the coefficients W_1 and W_0

$$-\frac{i}{4}(1-\xi^{-2})\sigma_3 W_1 = -\frac{\kappa}{2}\xi^{-2}(\sigma_1 \cos S - \xi\sigma_2 \sin S)W_0\,, \tag{A.49}$$

$$\frac{dW_0}{d\xi} = \left(\frac{1}{4\xi} + \frac{i\kappa^2}{1-\xi^2}\right)\sigma_3 W_0\,. \tag{A.50}$$

The latter equation has a solution of the form

$$W_0 = \begin{pmatrix} w & 0 \\ 0 & w^{-1} \end{pmatrix}, \quad w = \xi^{1/4}\left|\frac{1-\xi}{1+\xi}\right|^{-\frac{i\kappa^2}{2}}. \tag{A.51}$$

Hence we have

$$\hat{\Psi} = \begin{pmatrix} we^{i\theta} & -\varepsilon w^{-1}e^{-i\theta}\, 2i\kappa\dfrac{\cos S + i\xi\sin S}{\xi^2-1} \\ \varepsilon we^{i\theta}\cdot 2i\kappa\dfrac{\cos S - i\xi\sin S}{\xi^2-1} & w^{-1}e^{-i\theta} \end{pmatrix}. \tag{A.52}$$

Applying the reasoning similar to those of [32] it is easy to obtain from the normalization condition (A.16) the asymptotic expansion

$$\overset{\infty}{\Psi}(\xi,\varepsilon) = \hat{\Psi}(\xi,\varepsilon)\exp\sigma_3\left\{\mathcal{J}(\tau) - \frac{i\pi}{8} + \frac{3}{8}\ln 2\right\}\left[1 + O(\varepsilon^{1-\varkappa})\right], \tag{A.53}$$

where $\xi > 1 + \varepsilon^{\varkappa}$, $\frac{2}{3} < \varkappa < 1$. The remainder term here is evaluated by application of the same WKB-estimates used in Chapter 5 of the main text (see also [38]).

For the case $0 < \xi < 1$ the system (A.13) has to be transformed via (A.23) into the system (A.24), where $z = \xi^{-1}$. Taking into account the boundary conditions (A.18), (A.20) we have

$$H(\varepsilon)=i\sigma_3+O(\varepsilon)\ ,\quad \varepsilon\to 0\ ,$$

so that for $\xi<1-\varepsilon^{\varkappa}$, $\frac{2}{3}<\varkappa<1$ the following asymptotics takes place

$$\overset{\circ}{\Psi}(\xi,\varepsilon)=$$

$$=i\sigma_3\hat{\Psi}(\xi,\varepsilon)\exp\left\{\sigma_3\left[-\mathcal{I}(-\infty)+\frac{i\pi}{8}+\frac{3}{8}\ln 2\right]\right\}\left[I+O(\varepsilon^{1-\varkappa})\right]\ , \tag{A.54}$$

where $\mathcal{I}(-\infty)=-\frac{1}{8}\int\limits_{-\infty}^{-b}\tau^{-1}h^{-1}(\tau)\,d\tau$ and $\hat{\Psi}(\xi,\varepsilon)$ is defined by (A.52).

In order to calculate the connection matrix $Q=\overset{\circ}{\Psi}{}^{-1}\overset{\infty}{\Psi}$ it is necessary to expand any of the solutions (A.53), (A.54) beyond the boundary $|\xi|=1$. It is convenient to do this procedure assuming $\xi\in\mathbb{R}$ so that the exponents $\exp(\pm i\theta)$ in (A.53) remain to be bounded. The points $\xi=\pm1$ appear to be the turning points for the WKB-solution (A.52) as the representation (A.48) fails to be asymptotic in their neighbourhood. Thus a new representation would be constructed and then it would be matched with that of (A.53) and (A.54) as $\xi\to1$.

Introduce a new "stretched" variable in the neighbourhood of $\xi=1$,

$$\zeta=\varepsilon^{-1}(\xi-1)\ ,$$

and expand the right-hand side of the equation (A.46) in ζ , retaining only the leading order terms in ε :

$$\frac{d\tilde{\Psi}}{d\zeta}=\frac{1}{4}\begin{pmatrix} i\zeta & -2\kappa e^{+is} \\ -2\kappa e^{-is} & -i\zeta \end{pmatrix}\tilde{\Psi}\ . \tag{A.55}$$

The solution of this equation is expressed via the parabolic cylinder functions D_ν (see Chapter 5)

$$\tilde{\Psi}=\begin{pmatrix} D_{-\frac{i\kappa^2}{2}}\left(\frac{\zeta}{\sqrt{2i}}\right) & , \quad \frac{\kappa}{\sqrt{2i}}e^{is}D_{\frac{i\kappa^2}{2}-1}\left(\zeta\sqrt{\frac{i}{2}}\right) \\ \sqrt{\frac{i}{2}}\kappa e^{-is}D_{-\frac{i\kappa^2}{2}-1}\left(\frac{\zeta}{\sqrt{2i}}\right), & D_{\frac{i\kappa^2}{2}}\left(\frac{\sqrt{i}\,\zeta}{\sqrt{2}}\right) \end{pmatrix}. \tag{A.56}$$

A justification of the approximation (A.56) may be found in Chapter 5 of the main text where the WKB-estimates are derived in the neighbourhood of a turning point:

$$\overset{\infty}{\Psi}=\tilde{\Psi}\cdot C\cdot(I+O(\varepsilon^{3\text{æ}-2})),\ |\xi-1|<2\varepsilon^{\text{æ}},\ \frac{2}{3}<\text{æ}<1 .$$

Here we have a transitional domain $1+\varepsilon^{\text{æ}}<\xi<1+2\varepsilon^{\text{æ}}$, where the asymptotic matching of (A.56)' with (A.53) takes place. It may be achieved by choosing the right-hand constant matrix C as follows. The asymptotics of $\tilde{\Psi}$ as $\zeta\to+\infty$ is given by a well-known asymptotic expansions of the parabolic cylinder functions ([53]):

$$\tilde{\Psi}=\exp\left(\frac{i\zeta^2}{8}\sigma_3-\frac{\pi\kappa^2}{8}-\sigma_3\frac{i\kappa^2}{4}\ln(\zeta^2\cdot 2^{-1})\right)\left(I+O(\zeta^{-1})\right),\ \zeta\to+\infty , \tag{A.57}$$

while the asymptotics of $\overset{\infty}{\Psi}(\xi,\varepsilon)$ as $\xi\to 1+0$ has the form

$$\overset{\infty}{\Psi}=\exp\Big\{\Big[\frac{i}{4\varepsilon^2}+\frac{i(\xi-1)^2}{8\varepsilon^2}-\frac{i\pi}{8}+\mathcal{J}(\tau)-\frac{i\kappa^2}{2}\ln(\xi-1)+$$

$$+\left(\frac{3}{8}+\frac{i\kappa^2}{2}\right)\ln 2\Big]\sigma_3\Big\}\left(I+O(\varepsilon^{1-\text{æ}})+O\left(\frac{(\xi-1)^3}{\varepsilon^2}\right)\right). \tag{A.58}$$

Comparing these asymptotics one obtains the matching condition between $\tilde{\Psi}$ and $\overset{\infty}{\Psi}$ in the form

$$\overset{\infty}{\Psi} \sim \tilde{\Psi} \cdot C \quad \text{as} \quad \zeta \to +\infty, \; \xi \to 1-0,$$

where

$$C = e^{\frac{\pi\kappa^2}{8}} \cdot \exp\Big\{\sigma_3\Big[\frac{i}{4\varepsilon^2} - \frac{i\pi}{8} + \mathcal{J}(\tau) - \frac{i\kappa^2}{2}\ln\varepsilon +$$

$$+\Big(\frac{3}{8} + \frac{i\kappa^2}{4}\Big)\ln 2\Big]\Big\} \cdot \qquad \text{(A.59)}$$

Hence the representation (A.56) provides necessary expansion of $\overset{\infty}{\Psi}$ into the domain $\xi < 1$. Using again the asymptotics of the parabolic cylinder functions as $\zeta \to -\infty$ together with the formula (A.59), one derives the following expression for $\overset{\infty}{\Psi}$

$$\overset{\infty}{\Psi} = e^{\frac{\pi\kappa^2}{2}} \exp\Big\{\sigma_3\Big[\frac{i\zeta^2}{8} - \frac{i\kappa^2}{4}\ln(+2^{-1}\zeta^2) + \mathcal{J}(-\infty) + \frac{i}{4\varepsilon^2} - \frac{i\pi}{8}\Big]\Big\} \times$$

$$\times \begin{pmatrix} 1 & 2\sqrt{\pi}\,\dfrac{\exp\big(-\frac{\pi\kappa^2}{4} + iS - 2\mathcal{J}(-\infty)\big)}{\kappa\Gamma\big(-\frac{i\kappa^2}{2}\big)} \\ 2\sqrt{\pi}\,\dfrac{\exp\big(-\frac{\pi\kappa^2}{4} - iS + 2\mathcal{J}(-\infty)\big)}{\kappa\Gamma\big(\frac{i\kappa^2}{2}\big)} & 1 \end{pmatrix} \times \qquad \text{(A.60)}$$

$$\times \exp\Big\{\sigma_3\Big[\Big(-\frac{i\kappa^2}{2}\Big)\ln\varepsilon + \Big(\frac{3}{8} + \frac{i\kappa^2}{4}\Big)\ln 2\Big]\Big\}\Big(I + O(\varepsilon^{1-\varkappa}) + O\Big(\frac{(\xi-1)^3}{\varepsilon}\Big)\Big),$$

$$\varepsilon \to 0, \quad 2\varepsilon^{\varkappa} > 1 - \xi > \varepsilon^{\varkappa}.$$

In fact it may be rewritten in the variable $\xi = 1 + \varepsilon\zeta$, transforming (A.60) into an asymptotics of $\overset{\infty}{\Psi}$ as $\xi \to 1-0$. Now it may be compared with a similar asymptotics for $\overset{\circ}{\Psi}$, following from the representation (A.54)

$$\overset{\circ}{\Psi} = i\sigma_3 \exp\Big\{\sigma_3\Big[\frac{i}{4\varepsilon^2} + \frac{i(\xi-1)^2}{8\varepsilon^2} - \frac{i\pi}{8} - \mathcal{J}(-\infty) - \frac{i\kappa^2}{2}\ln\Big(\frac{\xi-1}{2}\Big) +$$

$$+\ln\varepsilon\Big]\Big\}\Big[I + O\Big(\frac{\varepsilon}{-\xi+1}\Big)\Big], \quad \varepsilon \to 0, \; \xi \to 1-0. \qquad \text{(A.61)}$$

The connection matrix $Q = \overset{\circ}{\Psi}{}^{-1} \overset{\infty}{\Psi}$ may be expressed directly from (A.60), (A.61). Its first coloumn has the form

$$Q_{11} = -ie^{\frac{\pi \kappa^2}{2} + 2\mathcal{J}(-\infty)} , \tag{A.62}$$

$$Q_{21} = \frac{\exp\left(-iS + \frac{i}{2\varepsilon^2} + \frac{\pi\kappa^2}{4}\right)}{\kappa\Gamma\left(\frac{i\kappa^2}{2}\right)} (2\pi^2)^{1/4} \cdot \left(\frac{\varepsilon^2}{2}\right)^{-\frac{i\kappa^2}{2}} \left(1 + O(\varepsilon^{\gamma})\right) \cdot \tag{A.63}$$

Remind that we have obtained these formulae under the assumption $\varepsilon \to 0$, i.e. $\tau \to -\infty$. Applying now the basic property of the monodromy matrix to be invariant in τ , one finds from (A.63) and (A.38) that

$$|Q_{21}|^2 = 2^{-3/2} |e^{\frac{\pi\kappa^2}{2}} - 1| = 2^{-3/2} \pi |a| ,$$

which implies finally

$$r(a) = \frac{1}{2} \kappa^2(a) = \frac{1}{\pi} \ln(1 - a\pi) . \tag{A.64}$$

Note, that $\kappa^2(a)$ may be negative according to our convention (A.44). Clearly the formula (A.64) implies $a < \frac{1}{\pi}$. Bringing this fact together with the estimates of the lemma 6 being uniform in a , we conclude that the upper boundary of the set L ,

$$L = \left\{ a \,\middle|\, \begin{array}{l} \Phi = \Phi(x,a) \text{ is increasing as } x \to \infty, \\ \exists x > 0, \ \Phi(x,a) > \frac{\pi}{2} \cdot \end{array} \right\},$$

coinsides with $\frac{1}{\pi}$.

Comparing the arguments of (A.63) and (A.38) we have

$$\Phi = \frac{S}{2} = x - r(a)\ln x + \frac{r(a)}{2}\ln 2 - 2\arg\Gamma\left(\frac{ir}{2}\right) + \pi l . \tag{A.65}$$

Note that for $a=0$ the initial problem (A.1), (A.2) has an explicit solution

$$\Phi(x,0)\equiv x .$$

Hence taking the limit $a\to 0$ in the formula (A.65) we get the values of the integer l :

$$l = \operatorname{sign} a$$

Therefore the statement A) of the Theorem is proved.

The statement B) is just a corollary of this result and that of the Lemma 1, where we have proved the estimate

$$\Phi'(x,\tfrac{1}{\pi})\geqslant 0 .$$

On the other hand, $\frac{1}{\pi}\notin L$ and the definition of the set L provides

$$\lim_{x\to\infty}\Phi(x,\tfrac{1}{\pi})=\frac{\pi}{2} ,$$

which proves the statement B) of the Theorem.

REMARK 1. Alongside with the proof of the Theorem the isomonodromic formulae provide an explicit calculation of the integral

$$J(\tau)=+\frac{1}{8}\int_{-b}^{\tau} t^{-1} h^{-1}(t)\,dt .$$

In fact, equating the expressions (A.38), (A.62) of the diagonal element Q_{11} , we have

$$e^{+\frac{\pi\tau}{2}}=\tau^{1/4}\exp\left\{+\frac{1}{4}\int_{-\infty}^{\tau} t^{-1} h^{-1}(t)\,dt\right\}\Bigg|_{\tau\to 0} , \qquad \text{(A.66)}$$

where the function $h(t)$ tends to 1 as $\tau\to 0$. The logarithmic singularity of the integral (A.66) is thus compensated by the multiplier $\tau^{-1/4}$.

§ 4. THE ASYMPTOTICS OF $\Phi(x,a)$ AS $x \to \infty$, FOR THE CASE $a > \frac{1}{\pi}$.

The remainder case C) of the Theorem would be treated in this paragraph. First we shall establish an existence of solution $\Phi(x,a)$ monotonically decreasing as $x \to \infty$ and then apply the method of isomonodromic deformations.

The proof of the asymptotics (A.3) for the solution $\Phi(x,a)$ is based on the following asymptotic formula, obtained in the recent paper [64] , [10] , [7] :

$$(\Phi'(x,\tfrac{2}{\pi})+1)^2 =$$

$$=4x^{-2}\sin^2\Phi(x,\tfrac{2}{\pi})\cdot\sin^2 x\left(1+O(x^{-1})\right),\quad x\to\infty . \tag{A.67}$$

We are going to prove now a similar expansion for all $a > \frac{1}{\pi}$.

The asymptotics (A.67) shows that $\Phi'(x,\frac{2}{\pi})<0$ for all $x>x^*$, and there exists a sequence $\{x_n\}$, $x_n > x^*$, such that $\Phi(x_n,\frac{2}{\pi})=-\pi n$, $n\geqslant m$. In the neighbourhood of each point the Taylor series expansion may be written in the form

$$\Phi(x) =$$

$$=-\pi n+x_n-x-\frac{1}{x_n}(x-x_n)^2+b_n(x-x_n)^3+O[(x-x_n)^4], \tag{A.68}$$

$$\Phi'(x)=-1-\frac{2}{x_n}(x-x_n)+3b_n(x-x_n)^2+O[(x-x_n)^3],\quad x\to x_n .$$

Introduce the function $v(x)=x(1+\Phi'(x,\frac{2}{\pi}))$ and consider an integral equation for it, which is equivalent to initial equation (A.1). Performing the transformations similar to (A.42), (A.43) we obtain this equation in the form

$$v = c_n \sin^2\Phi + \sin 2\Phi + \frac{v^2}{2x} + \sin^2\Phi \int_{x_n}^{x} \frac{v^2(y)\,dy}{2y^2 \sin^2\Phi(y)} . \tag{A.69}$$

Applying the representation (A.67) it may be presented as follows

$$v = \kappa \cdot \sin^2\Phi + \sin 2\Phi + \frac{v^2}{2x} - \sin^2 \Phi(x) \int_{x}^{\infty} \frac{v^2(y)\,dy}{2y^2 \sin^2\Phi(y)} , \tag{A.70}$$

where

$$\kappa = c_n + \int_{x_n}^{\infty} \frac{v^2(y)\,dy}{2y^2 \sin^2\Phi(y)} .$$

Hence we have a new form of asymptotic representation (A.67)

$$\Phi'(x,a) = -1 + \frac{1}{x}\left[\sin 2\Phi + \kappa(a)\sin^2\Phi\right] + O(x^{-2}),\ x \to \infty,\ a = \frac{2}{\pi} . \tag{A.71}$$

Introduce the set Δ consisting of all $a > \frac{1}{\pi}$ such that $\Phi(x,a)$ has the asymptotic expansion (A.71) as $x \to \infty$. It is non-empty since $\frac{2}{\pi} \in \Delta$.

LEMMA 8. The set Δ is open. If $x > 2|\kappa(a)| + 4$, $a \in \Delta$, then $\Phi'(x,a) < 0$.

PROOF. Let $a_0 \in \Delta$, then the formula (A.71) takes place for $a = a_0$. It implies the existence of $x^*(a_0) > 1$ such that $\Phi'(x,a_0) < 0$ for all $x > x^*$. Consider an increasing sequence of points $x_n > x^*$ for which $\Phi(x_n, a_0) = -\pi n, n \geqslant m$. Applying the expansion (A.68) to the right-hand sides of (A.70) and (A.69), we have

$$3b_n(a_0)x_n(a_0) = \kappa(a_0) + O(x_n^{-1}(a_0)) , \tag{A.72}$$

$$3b_n(a_0)x_n(a_0) = c_n(a_0) + \frac{2}{x_n(a_0)}, \quad n \to \infty.$$

Let us fix $n = N$ to be large enough to provide the estimate

$$2|C_N(a_0)| + 4 < \frac{1}{100} x_N(a_0). \tag{A.73}$$

Assume the parameter a to lie sufficiently close to a_0. According to lemma 1 we have that for all $x > x_N(a)$ such that $\Phi'(x,a) < 0$ the function $v(x,a) = x[1 + \Phi'(x,a)]$ satisfies the integral equation (A.69). Hence the inequality (A.73) holds for all a being sufficiently close to a_0.

Consider the function $g(x,a) = |v(x,a)\sin^{-1}\Phi(x,a)|$ for $x > x_N(a)$, $\Phi'(x,a) < 0$. The estimate $v(x,a) < x$ together with equation (A.69) yield

$$g(x,a) \leq 2C_N(a) + 4 + \int_{x_N}^{x} \frac{g^2(y)}{y^2}\, dy.$$

Applying finally the Bihari's lemma ([69]) to this integral inequality we obtain the estimate

$$g(x,a) \leq \frac{2|C_N(a)| + 4}{1 - x_N^{-1}(a)(2|C_N(a)| + 4)} < x_N(a) \tag{A.74}$$

for all $x > x_N(a)$. Hence $v(x,a) < x$ for all $x > x_N(a)$, so that the integral equations (A.70), (A.69) provide $a \in \Delta$, which proves that Δ is open.

The second statement of the lemma follows immediatly from the estimates

$$\frac{1}{2} x^*(a) > \sin 2\Phi(x^*, a) \quad \text{for } C_N(a) < 0,$$
$$\frac{1}{2} x^*(a) > K(a) \sin^2\Phi(x^*, a) + \sin 2\Phi(x^*, a) \quad \text{for } C_N(a) > 0,$$

where x^* is the point in which $\Phi'(x^*,a)=0$. The lemma is proved.

We proceed now to the isomonodromic deformation method in order to calculate the connection matrix Q for the system (A.13) as $x\to\infty$. The existence of the solution $\Phi(x,a)$ which has an asymptotics as $x\to\infty$ of the form (A.3) is proved above, so we substitute it via the transformation (A.12) into the coefficients of (A.13). Denote again

$$S=\frac{1}{2}\Phi\,,\quad \varepsilon=x^{-1}\,,\quad \lambda=\varepsilon\xi\,,$$

and rewrite the system (A.13) in the form:

$$\varepsilon\frac{d\Psi}{d\xi}=\Big[\frac{i}{8}\sigma_3-\frac{i\sigma_1}{4\xi\sin S}+\frac{1}{8\xi^2}\Big(\frac{2\cos S}{\sin^2 S}\sigma_1+$$

$$+i\frac{1+\cos^2 S}{\sin^2 S}\sigma_3\Big)\Big]\Psi+\frac{O(\varepsilon^2)}{\xi^2}\begin{pmatrix} O(1)\,, & O(\xi) \\ O(\xi)\,, & O(1)\end{pmatrix}\Psi \qquad \text{(A.75)}$$

The forthcoming calculations follow closely to those accomplished in § 3. Therefore we scetch briefly the main points of the asymptotic procedure omitting rather cumbersome details.

The WKB-ansatz for the solution of the system (A.75) is taken in the form

$$\hat{\Psi}=\big[-\xi\sin S+i\sigma_3\cos S+\sigma_1\big]\frac{(\sin S)^{-1}}{\sqrt{\xi^2-1}}\times$$

$$\times\Big[W_0(\xi)e^{i\theta\sigma_3}+\varepsilon W_1(\xi)e^{-i\theta\sigma_3}\Big] \qquad \text{(A.76)}$$

$$\theta=\frac{1}{8\varepsilon}(\xi+\xi^{-1}),\qquad \varepsilon\to 0\,,\quad |\xi|\neq 1\,,$$

Substituting it into the system (A.75) and equating the terms of orders $O(1)$ and $O(\varepsilon)$ we get the first order equations on W_0, W_1 similar to those of (A.49), (A.50). The constant of integration for W_0 is determined via the boundary value conditions (A.16), (A.18). A justification of the WKB-approximation (A.76) is proved in a standard way (see [38]).

The points $\xi = \pm 1$ are again the turning points of the system (A.75). The asymptotic expansions of the solutions $\overset{\circ}{\Psi}, \overset{\infty}{\Psi}$ in the neighbourhood of $\xi = 1$ have the form

$$\overset{\infty}{\Psi} = \left[i\sigma_3 e^{iS\sigma_3} + \sigma_1 \right] \frac{\sin^{-1} S}{\sqrt{2(\xi-1)}} \exp\left\{ \sigma_3 \left[\frac{i}{4\varepsilon} + \frac{i(\xi-1)^2}{8\varepsilon} + \mathcal{J}(\tau) + \right.\right.$$

$$\left.\left. + \frac{i\pi}{8} - \frac{iK}{4} \ln(\xi-1) + \left(\frac{iK}{4} - \frac{3}{8} \right) \ln 2 \right\} \left[I + O\left(\frac{\varepsilon}{\xi-1} \right) + \right.$$

$$\left. + O(\xi - 1 + \varepsilon^{-1}(\xi-1)^3) \right], \quad \varepsilon \to 0, \ \xi \to 1+0, \ \varepsilon^{\varkappa} < |1-\xi| < 2\varepsilon^{\varkappa} \tag{A.77}$$

$$\overset{\circ}{\Psi} = \left[+i\sigma_3 \cos S + \sigma_1 \right] \left[i\sigma_3 e^{iS\sigma_3} + \sigma_1 \right] \frac{\sin^{-2} S}{\sqrt{2(1-\xi)}} \exp\left\{ \sigma_3 \left[\frac{i}{4\varepsilon} + O\left(\frac{\varepsilon}{(\xi-1)} \right) + \right.\right.$$

$$\left.\left. + \frac{i(\xi-1)^2}{8\varepsilon} - \mathcal{J}(\tau) + \frac{i\pi}{8} - \frac{iK}{4} \ln(1-\xi) + \left(\frac{iK}{4} - \frac{3}{8} \right) \ln 2 \right\} \left[I + O\left(\frac{\varepsilon}{\xi-1} \right) + \right.$$

$$\left. + O(1 - \xi + \varepsilon^{-1}(1-\xi)^3) \right], \quad \varepsilon \to 0, \ \xi \to 1-0, \ \varepsilon^{\varkappa} < 1-\xi < 2\varepsilon^{\varkappa} \tag{A.78}$$

where $\frac{1}{3} < \varkappa < \frac{1}{2}$.

In order to calculate the connection matrix it is necessary to expand the asymptotics (A.77) into the domain $\xi < 1$. This procedure repeats by word the similar calculations of § 3, involving the matching via the parabolic cylinder functions. As a result we have finally the explicit firmulae for the coloumn of the connection matrix

$$Q = \overset{\circ}{\Psi}{}^{-1}\overset{\infty}{\Psi}:$$

$$Q_{11} = e^{\frac{\pi\kappa}{4} + 2\mathfrak{I}(-\infty)}, \tag{A.79}$$

$$Q_{21} = \sqrt{\pi}\,\frac{\exp\{iS + \frac{i}{2\varepsilon} + \frac{i\pi}{4} + \frac{\pi\kappa}{8}\}}{\Gamma(\frac{1}{2} - \frac{i\kappa}{4})} 2^{\frac{i\kappa}{4} - \frac{1}{4}} \varepsilon^{-\frac{i\kappa}{4}}, \quad \varepsilon \to 0. \tag{A.80}$$

The isomonodromic condition provides an invariance of Q in $x = \varepsilon^{-1}$, so that its elements (A.79), (A.80) have to coinside with that of (A.38). Comparing the non-diagonal functions Q_{21}, we get

$$|Q_{21}|^2 = 2^{-3/2}\left(e^{\frac{\pi\kappa}{2}} + 1\right) = 2^{-3/2}\pi|a|,$$

which yields

$$r(a) = \frac{\kappa(a)}{2} = \frac{1}{\pi}\cdot \ln(a\pi - 1), \quad a > \frac{1}{\pi}. \tag{A.81}$$

Equating the arguments in (A.80), (A.38) we obtain the formula (A.8).

To end the proof of the Theorem it remains to show that every $a > \frac{1}{\pi}$ belongs to Δ.

LEMMA 9. Let $a_0 > \frac{1}{\pi}$ and $a_0 \in \partial\Delta$. Then $a_0 \in \Delta$.

PROOF. The result of the lemma 8 together with the constraint (A.81) yield the existence of $x^* > 0$ such that $\Phi'(x,a) < 0$ for all $x > x^*$. Hence applying the lemma 1 we have $\Phi'(x,a_0) < 0$ for $x > x^*$, and the integral equation (A.70) provides the estimate

$$v(x,a_0) \leq \kappa(a_0)\sin^2\Phi(x,a_0) + \sin 2\Phi(x,a_0) + \frac{v^2(x,a_0)}{2x},$$

$$x > x^*,$$

where $v = x(\Phi' + 1)$. It implies the existence of x_*, such that $\Phi'(x_*, a_0) < -\delta < 0$ for $x > x_*$. Turning now to the integral equation (A.70) we assume all $c_n \leq 0$, because for any $c_N > 0$ the estimate (A.74) holds and hence $a_0 \in \Delta$ immediately. If all $c_n \leq 0$, then the integral in (A.70) remains bounded as $x \to \infty$ and hence the asymptotics (A.71) holds, which means that $a_0 \in \Delta$. The lemma is proved.

Combining the results of the lemmas 8 and 9 and the fact of non-emptiness of $\Delta\ (\frac{2}{\pi} \in \Delta)$, we conclude that Δ coinsides with the set $a > \frac{1}{\pi}$. The statement C) of the Theorem is proved completely.

The author wishes to thank V.Yu. Novokshenov for the formulation of the problem. He also thanks A.M.Il'in for some useful consultations on qualitative analysis of nonlinear equations.

APPENDIX 2

SINGULAR SOLUTIONS OF THE PAINLEVÉ II EQUATION

by A.A.Kapaev

We discuss here the real-valued singular solutions of the equation

$$u_{xx} - xu - 2u^3 = 0 \qquad \text{(A.2.1)}$$

and the monodromy data associated with corresponding linear system (1.9).

1. Since the classical work [1] by P.Painlevé it is known that all solutions of the equation (A2.1) are meromorfic in x, $x \in \mathbb{C}$. The Laurent series in the neighbourhood of any pole x_n has the form

$$u(x) = a_{-1}(x - x_n)^{-1} + \sum_{k=1}^{\infty} a_k (x - x_k)^k =$$

$$= a_{-1}\left[(x - x_n)^{-1} - \frac{1}{6} x_n (x - x_n) - \frac{1}{4}(x - x_n)^2 + d_n (x - x_n)^3 + \ldots\right], \qquad \text{(A2.2)}$$

where all the coefficients a_k are uniquely determined via the coordinate of pole x_n, the residue $a_{-1} = \pm 1$ and the coefficient $d_n = a_{-1} a_3$.

The main purpose of this Appendix consists of a calculation of pole's coordinates x_n together with the values of d_n in terms of corresponding monodromy data p, q. We assume throughout the text that

$$\operatorname{Im} x_n = 0, \quad |x_n| \to \infty, \quad n \to \infty, \quad a_{-1} = +1^{*)}.$$

---

*) It is quite sufficient to study the case $a_{-1} = 1$. The opposite case $a_{-1} = -1$ may be obtained by changing signs $u \mapsto -u \Leftrightarrow p \mapsto -p$, $q \mapsto -q$, $r \mapsto -r$.

It is convenient to introduce the new function

$$\Phi(\lambda) = (I - i\sigma_1)\Psi(\lambda) , \qquad (A2.3)$$

where Ψ is a solution of the system (0.9). We transform it into the second order scalar equation on the first-line elements of the matrix Φ :

$$\frac{\partial}{\partial\lambda}\left(P_0 \frac{\partial}{\partial\lambda} y\right) + P_0 (Q_0 + R_0) y = 0 , \qquad (A2.4)$$

where $y = (\Phi_{11}, \Phi_{12})$,

$$P_0 = [\lambda^2 + (x - x_n)^{-2} + \frac{1}{6} x_n + O(x - x_n)]^{-1} ,$$

$$Q_0 = (4\lambda + x)^2 + 40 d_n - \frac{7}{9} x_n^2 + O[x_n (x - x_n)] ,$$

$$R_0 = -8\lambda^2 \{(x - x_n)^{-1} + O[x_n (x - x_n)]\} P_0 .$$

We have substituted here the Laurent expansion (A2.2) instead of the function $u(x)$ in (1.9). The reason for transformation (A2.3) and transition to the scalar equation (A2.4) consists of the necessary to obtain a differential equation on the Ψ-function with non-singular coefficients as $x \to x_n$.

The canonical solutions of the equation (A2.4) are fixed by the condition

$$y_k(\lambda) \to \left(e^{-i(\frac{4}{3}\lambda^3 + x\lambda)} , -ie^{i(\frac{4}{3}\lambda^3 + x\lambda)}\right),$$

$$\lambda \in \Omega_k , \quad \lambda \to \infty . \qquad (A2.5)$$

The Stokes matrices S_k are defined as usual:

$$y_{k+1}(\lambda) = y_k(\lambda) S_k , \ k = 1, 2, \ldots, 6 . \qquad (A2.6)$$

Obviously they coinside with corresponding Stokes matrices for the

Ψ-function, through they can not be expressed explicitly through the parameters x, x_n, d_n . Nevertheless their asymptotics as $x_n \to \pm\infty$, $|x - x_n| \leq O(e^{-|x_n|})$ may be calculated quite effectively. As a result one obtains an asymptotic distribution of poles x_n in terms of parameters p and q . The similar formulae were obtained in the main text of the book (Chapter 10) for the case $x_n \to +\infty$. Here we present a new proof of this result, and besides our method here provides the treatment of the case $x_n \to -\infty$.

The results of this Appendix just complete the analysis of real-valued solutions of the equation (A1.1). In terms of the monodromy data the only case we are missing here is described by the condition $|p| = 1$, which extracts one-parameter submanifold of solutions (see the Chapter 1). Concerning the asymptotic description of this submanifold, we propose a certain hypothesis at the end of the text.

2. Consider first the case $x_n \to -\infty$. Having the large parameter x_n , we treat the equation (A2.4) with the help of the WKB-method. Rescaling the variable λ via the transform

$$\lambda = \frac{1}{2}\sqrt{-x_n}\; z\,, \qquad \text{(A2.6')}$$

we obtain the equation

$$\frac{\partial}{\partial z}\left(P \frac{\partial}{\partial z} y\right) + \frac{(-x_n)^3}{4} P(Q + R)y = 0$$

$$P = [z^2 + 4(-x_n)^{-1}(x - x_n)^{-2} - \frac{2}{3} + O((-x_n)^{-1}(x - x_n))]^{-1},$$

$$Q = [z^2 - 1 + (-x_n)^{-1}(x - x_n)]^2 + 40 d_n x_n^{-2} - \frac{7}{9} + O((-x_n)^{-1}(x - x_n)),$$

$$R = -8z^2\,[(x - x_n)^{-1} + O(x_n(x - x_n))]\,P x_n^{-2}.$$

The canonical solutions have the following asymptotics in the variable z :

$$y_k(z) \to \left(\exp\left[-\frac{i}{2}(-x_n)^{3/2}\left(\frac{z^3}{3} - z\right)\right], -i\exp\left[\frac{i}{2}(-x_n)^{3/2}\left(\frac{z^3}{3} - z\right)\right]\right). \qquad \text{(A2.7')}$$

In order to get the same structure of the Stokes lines as in the case of regular solutions of PIII equation (A2.1) (see Chapter 9), we assume that

$$|40 d_n x_n^{-2} - \frac{7}{9}| \leq O((-x_n)^{-3/2}), \quad x_n \to -\infty . \tag{A2.8}$$

The turning points for the equation (A2.7) are expressed as follows

$$z_{1,2,3,4} = \pm 1 + O((-x_n)^{-3/4}), \tag{A2.9}$$

$$z_{5,6,7,8} = \pm 2i(-x_n)^{-1/2}(x-x_n)[1+O(x_n(x-x_n)^2)] . \tag{A2.9'}$$

The points $z_{5,6}$ and $z_{7,8}$ are of order $+1$ and -1 respectively and they are associated with the poles of the function $R(z)$. The conjugate Stokes lines are determined by the equations

$$\operatorname{Re} i \int_{z_k}^{z} \sqrt{Q(\tau) + R(\tau)}\, d\tau = 0 .$$

They tend at infinity to the rays $\arg z = \frac{\pi k}{3}$, $k=1,2,\ldots,6$. The corresponding picture of the conjugate Stokes lines is presented at the fig.A.1. For the WKB-approximations of the solutions of (A2.7) the usual formulae take place

$$y_{WKB} = \left[P^2(Q+R)\right]^{-1/4} \left(\exp\left[-\frac{i}{2}(-x_n)^{3/2} \int_{z_0}^{z} \sqrt{Q+R}\, d\tau\right], \right.$$

$$\left. \exp\left[\frac{i}{2}(-x_n)^{3/2} \int_{z_0}^{z} \sqrt{Q+R}\, d\tau\right]\right) . \tag{A2.10}$$

The reasoning quite similar to that of [38] yields the WKB-estimates in the domains D_k, shown at the fig. A.2, where the neighbourhoods of diameter $O((-x_n)^{-3/4+\delta})$ around the turning points are removed,

$$y(z)=y_{WKB}(z)\left[1+O((-x_n)^{-2\delta})\right],\ z\in D_\kappa \qquad \text{(A2.11)}$$

$$y(z)=y_{WKB}(z)\left[1+O(z^{-1}(-x_n)^{-3/2})\right],\ z\in D_\kappa,\quad z\to\infty , \qquad \text{(A2.11')}$$

where $y(z)$ is some suitable solution of the equation (A2.7).

As the domains D_κ are mapped uniquely into the canonical sectors Ω_κ, the solutions y_{WKB} differ from the canonical solutions y_κ only by diagonal right-hand matrix multipliers which are independent of z :

$$y_\kappa(z)=y_{WKB_\kappa}(z)\left[1+O((-x_n)^{-2\delta})\right]C_\kappa ,$$
$$z\in D_\kappa ,\quad x_n\to-\infty . \qquad \text{(A2.12)}$$

Similarly to the case of regular solution of PII equation it is sufficient to calculate two of the Stokes matrices – S_6 and S_1 . Let us put $z_0=1$ in equation (A2.10) and fix the branches of functions $[P^2(Q+R)]^{-1/4}$ and $\sqrt{Q+R}$ by the conditions

$$[P^2(Q+R)]^{-1/4}\longrightarrow 1 ,\qquad z\longrightarrow+\infty , \qquad \text{(A2.13)}$$

$$\sqrt{Q+R}\longrightarrow z^2 ,\qquad z\longrightarrow+\infty . \qquad \text{(A2.13')}$$

The phase integral

$$\int_1^z \sqrt{Q(\tau)+R(\tau)}\,d\tau$$

is calculated by the usual asymptotic procedure as $z\to\infty ,\ z\in \mathcal{D}_\kappa$ (see Chapter 5). As a result we have

$$C_\kappa = e^{\varphi\sigma_3}\begin{pmatrix}1 & 0\\ 0 & -i\end{pmatrix},\quad \kappa=1,2,6 , \qquad \text{(A2.14)}$$

where

$$\varphi=\frac{i}{2}(-x_n)^{3/2}\left[\frac{2}{3}+\frac{c_n}{8}+\frac{3c_n}{4}\ln 2-\frac{c_n}{8}\ln c_n+o((-x_n)^{-3/2})\right],$$

$$c_n=40d_nx_n^{-2}-\frac{7}{9}=O((-x_n)^{-3/2}).$$

Thus we have constructed the solution of the equation (A2.7) outwards the neighbourhood of the turning points z_k. Let us proceed now to the neighbourhood with a diameter of $O((-x)^{-3/4+\delta_1})$, $\delta_1\geqslant\delta$, surrounding the turning points z_1, z_2. The equation (A2.7) there becomes much simpler:

$$\frac{d^2y_0}{dz^2}+\frac{1}{4}(-x_n)^3\left[4(z^2-1)+c_n\right]y_0=0. \tag{A2.15}$$

The change of variables

$$z-1=e^{-i\pi/4}\frac{1}{\sqrt{2}}(-x_n)^{-3/4}\zeta$$

reduces the equation (A2.15) to the well-known Weber-Hermite equation

$$\frac{d^2y_0}{d\zeta^2}+\left(\nu+\frac{1}{2}-\frac{\zeta^2}{4}\right)y_0=0, \tag{A2.16}$$

where $\nu+\frac{1}{2}=e^{-i\pi/2}(-x_n)^{3/2}\frac{c_n}{8}$, and the solutions of which are expressed in terms of Weber-Hermite functions

$$y_0(\zeta)=\left(D_\nu(\zeta),\ D_{-\nu-1}(i\zeta)\right). \tag{A2.17}$$

Evaluating the right-hand side of the equation (A2.7) for the solution (A2.17) it is easy to obtain the estimate

$$|y(z)-y_0(z)|\leqslant O((-x_n)^{-4\delta_1}) \tag{A2.18}$$

as $|z-1|\leqslant O((-x_n)^{-9/16-\delta_1})$. Comparing it with (A2.11) one concludes that the solutions y_{WKB} and y_0 are matched with each other as $|z-1| = O((-x_n)^{-5/8})$, while their difference does not exceed $O((-x_n)^{-1/4})$. In this domain $|\zeta| \to \infty$, hence y_0 may be replaced by its asymptotics:

$$y_0(\zeta) = \left(e^{-\zeta^2/4 + \nu \ln \zeta}, e^{\zeta^2/4 - (\nu+1)\ln \zeta}\right)\left[1 + O(\zeta^{-2})\right]\begin{pmatrix} 1 & 0 \\ 0 & e^{-i\pi/2(\nu+1)} \end{pmatrix},$$

$$\zeta \to \infty, \quad \arg \zeta = -\frac{\pi}{4}, \qquad \text{(A2.19)}$$

$$y_0(\zeta) = \left(e^{-\zeta^2/4 + \nu \ln \zeta}, e^{\zeta^2/4 - (\nu+1)\ln \zeta}\right)\left[1 + O(\zeta^{-2})\right] \times$$

$$\times \begin{pmatrix} 1 & \dfrac{\sqrt{2\pi}}{\Gamma(\nu+1)} e^{-i\frac{\pi}{2}\nu} \\ \dfrac{\sqrt{2\pi}}{\Gamma(-\nu)} e^{i\pi(\nu+1)} & e^{i\frac{3\pi}{2}(\nu+1)} \end{pmatrix}, \quad \zeta \to \infty, \quad \arg \zeta = \frac{3\pi}{4}.$$

In the same domain we evaluate again the phase integral, applying the same methods used above (see Chapter 5). Expressing its asymptotics in the variable ζ, we have

$$y_{WKB} =$$

$$= A \cdot \left(e^{-\zeta^2/4 + \nu \ln \zeta}, e^{\zeta^2/4 - (\nu+1)\ln \zeta}\right)\left[1 + O(\zeta^{-2})\right] \exp(\sigma_3 \psi), \qquad \text{(A2.20)}$$

where

$$\psi = \frac{1}{2}\left(\nu + \frac{1}{2}\right) - \frac{1}{2}\left(\nu + \frac{1}{2}\right)\ln\left(\nu + \frac{1}{2}\right) - i\frac{\pi}{2}\left(\nu + \frac{1}{2}\right),$$

$$A = 2^{3/4} e^{i\pi/8} (-x_n)^{-1/8} (x - x_n)^{-1}.$$

Therefore the asymptotics (A2.19), (A2.20) yield the matching conditions of the form

$$y_{WKB_k}(\zeta) = A y_0(\zeta)\left[1 + O(\zeta^{-2})\right] N_k \ , \quad k = 2, 6 \ , \tag{A2.21}$$

where the matrices N_k have the explicit expressions

$$N_2 = \begin{pmatrix} e^{2\pi i(\nu+1)} \ , & -\frac{\sqrt{2\pi}}{\Gamma(\nu+1)} e^{i\pi/2} \\ -\frac{\sqrt{2\pi}}{\Gamma(-\nu)} e^{3i\pi/2(\nu+1)} \ , & e^{i\pi/2(\nu+1)} \end{pmatrix}, \ exp(\psi\sigma_3) \tag{A2.22}$$

$$N_6 = \begin{pmatrix} 1 & 0 \\ 0 & e^{i\pi/2(\nu+1)} \end{pmatrix} exp(\psi\sigma_3) \ .$$

Bringing together the formulae (A2.14) and (A2.22) we obtain the explicit expressions for the Stokes matrices:

$$S_6 S_1 = \begin{pmatrix} 1-pq & -q \\ p & 1 \end{pmatrix} \simeq C_6^{-1} N_6^{-1} N_2 C_2 =$$

$$= \begin{pmatrix} e^{2\pi i(\nu+1)} \ , & -\frac{\sqrt{2\pi}}{\Gamma(\nu+1)} e^{-f} \\ \frac{\sqrt{2\pi}}{\Gamma(-\nu)} e^{i\pi(\nu+\frac{1}{2})+f} \ , & 1 \end{pmatrix} , \tag{A2.23}$$

where

$$f = 2(\varphi+\psi) = i\frac{2}{3}(-x_n)^{3/2} - \frac{3}{2}(\nu+\tfrac{1}{2})\ln(-x_n) - 3(\nu+\tfrac{1}{2})\ln 2 - i\frac{\pi}{2}(\nu+\tfrac{1}{2}) \ .$$

Applying finally the double argument formula for the Γ-function we have

$$\left.\begin{aligned} p &= \frac{1}{\sqrt{2}} \frac{\Gamma(-\nu-\frac{1}{2})}{\Gamma(-2\nu-1)} e^{i\pi/2(\nu+1/2)} e^{h} , \\ q &= \frac{1}{\sqrt{2}} \frac{\Gamma(\nu+\frac{1}{2})}{\Gamma(2\nu+1)} e^{i\pi/2(\nu+1/2)} e^{-h} \end{aligned}\right\} , \tag{A2.24}$$

$$h = i\frac{2}{3}(-x_n)^{3/2} - \frac{3}{2}(\nu+\frac{1}{2})\ln(-x_n) - 5(\nu+\frac{1}{2})\ln 2 + i\frac{\pi}{2}(1-a_{-1}) .$$

The isomonodromic condition, i.e. independence of the monodromy data p and q of x, implies the independence of x_n, because we apply the theorem of Flaschka and Newell (see Theorem 3.1 of the main text) to the values of x tending to x_n but not coinsiding with x_n, where the Painlevé function $u(x)$ has the pole.

The fact of independence of x_n in the formulae (A2.24) yield immediately, that

$$\nu + \frac{1}{2} \in i\mathbb{R} ,$$

which implies

$$\begin{aligned} p = \bar{q} , \quad pq = |p|^2 = 1 + e^{2\pi i(\nu+1/2)} > 1 , \\ r = -\frac{2\operatorname{Re} p}{|p|^2-1} = \bar{r} . \end{aligned} \tag{A2.25}$$

Thus we have reestablished the real-valued reduction $p=\bar{q}$, of the monodromy data directly from the isomonodromic condition and initial assumption $\operatorname{Im} x_n = 0$.

The formulae (A2.24) turn to be now the equations with respect to x_n. Clearly, they have an infinite set of solutions due to the multivalued argument in the exponents. Finally the leading term for the asymptotic distribution of poles of the Painlevé function (A2.1) has the form:

$$(-x_n)^{3/2} =$$

$$= 3\pi n - \frac{3\alpha}{2}\ln 3\pi n - \frac{9}{2}\alpha \ln 2 + \frac{3}{2}\arg\Gamma\left(\frac{1}{2}+i\alpha\right) + \frac{3}{2}\theta - \frac{3\pi}{4}(1-a_{-1}), \quad \text{(A2.26)}$$

where

$$\alpha = \frac{1}{2\pi}\ln(|p|^2-1), \quad \theta = \arg p, \quad n \to +\infty.$$

Note that the asymptotics (A2.26) may be rewritten in the form

$$(-x_n)^{3/2} =$$

$$= \frac{3\pi}{2}n - \frac{3\alpha}{2}\ln\frac{3\pi}{2}n - \frac{9\alpha}{2}\ln 2 + \frac{3}{2}\arg\Gamma\left(\frac{1}{2}+i\alpha\right) + \frac{3}{2}\theta, \quad \text{(A2.26')}$$

where even values of n are associated with positive residues $a_{-1} = +1$ and odd values of n - with negative residues $a_{-1} = -1$. It means that poles x_n with the residues $+1$ and -1 alternate each other beginning from sufficiently large number of n.

3. We proceed now to the case $x_n \to +\infty$. The change of variables

$$\lambda = \frac{1}{2}\sqrt{x_n}\, z$$

reduces the equation (A2.4) to the following one

$$\frac{\partial}{\partial z}\left(P\frac{\partial}{\partial z}y\right) + \frac{1}{4}x_n^2 P(Q+R)y = 0, \quad \text{(A2.27)}$$

where

$$P = \left[z^2 + 4x_n^{-1}(x-x_n)^{-2} + \frac{2}{3} + O(x_n^{-1}(x-x_n))\right]^{-1},$$

$$Q = z^4 + 2z^2\left[1 + x_n^{-1}(x-x_n)\right] + 40 d_n x_n^{-2} + \frac{2}{9} + O(x_n^{-1}(x-x_n)),$$

$$R = -8z^2\left[(x-x_n)^{-1} + O(x_n(x-x_n))\right]P x_n^{-2}.$$

The canonical solutions of the equation (A2.27) have the following asymptotics

$$y_K(z) \to \left(\exp\left[-\frac{i}{2}x_n^{3/2}\left(\frac{1}{3}z^3+z\right)\right], -i\exp\left[\frac{i}{2}x_n^{3/2}\left(\frac{1}{3}z^3+z\right)\right]\right),$$
$$z \in \Omega_K, \quad z \to \infty \quad . \tag{A2.28}$$

Just as above we assume that

$$\left|40 d_n x_n^{-2} + \frac{2}{9}\right| \leq O(x_n^{-3/2}), \quad x_n \to +\infty . \tag{A2.29}$$

This estimate provides the coinsidence of the Stokes lines and turning points, shown at the fig. 3, with those associated with regular solutions of PII equation (A2.1):

$$z_{1,2} = O(x_n^{-3/4}) \quad ,$$
$$z_{3,4} = \pm i\sqrt{2} + O(x_n^{-3/2}), \tag{A2.30}$$
$$z_{5,6,7,8} = \pm i 2 x_n^{-1/2}(x-x_n)^{-1}\left[1+O(x_n(x-x_n)^2)\right] .$$

The points $z_{5,6}$ and $z_{7,8}$ are generated by the poles of coefficient $R(z)$ and have the order of $+1$ and -1 respectively. We define the WKB-solutions as usual:

$$y_{WKB} = \left[P^2(Q+R)\right]^{-1/4}\left(\exp\left[-\frac{i}{2}x_n^{3/2}\int_{z_0}^{z}\sqrt{Q+R}\,d\tau\right],\right.$$
$$\left.\exp\left[\frac{i}{2}x_n^{3/2}\int_{z_0}^{z}\sqrt{Q+R}\,d\tau\right]\right) . \tag{A2.31}$$

The following estimates take place in the domains D_K, shown at fig. A.4, where the neighbourhoods of diameter $O(x_n^{-3/4+\delta})$ around the turning point are removed

$$y(z) = y_{WKB}(z)\left[1 + O(x_n^{-2\delta})\right], \quad z \in D_\kappa ,$$
(A2.32)
$$y(z) = y_{WKB}(z)\left[1 + O(x_n^{-3/2} z^{-1})\right], \quad z \in D_\kappa, \quad z \to \infty ,$$

where $y(z)$ is an exact solution of the equation (A2.27).

Because of the mutual disposition of domains D_κ it is most convenient to calculate the product of the Stokes matrices $S_1 S_2 S_3$. Let us put $z_0 = 0$ in (A2.31) and fix the branches of functions $[P^2(Q+R)]^{-1/4}$ and $\sqrt{Q+R}$ by asymptotic conditions

$$[P^2(Q+R)]^{-1/4} \longrightarrow 1, \quad z \longrightarrow \infty ,$$
$$[P^2(Q+R)]^{-1/4} \longrightarrow 1, \quad z \longrightarrow -\infty , \tag{A2.33}$$
$$\sqrt{Q+R} \longrightarrow z^2 , \qquad z \longrightarrow \pm\infty .$$

The calculation of the phase integral

$$\int_0^z \sqrt{Q+R}\, d\tau$$

as $z \longrightarrow \pm\infty$ proceeds just as for the regular solution case (see Chapter 5). In the domains D_1 and D_4 we have

$$y_{WKB_1}(z) = \left(\exp\left[-\frac{i}{2} x_n^{3/2}\left(\frac{1}{3}z^3 + z\right)\right], \exp\left[\frac{i}{2} x_n^{3/2}\left(\frac{1}{3}z^3 + z\right)\right]\right)\exp(-\varphi\sigma_3),$$

$$y_{WKB_4}(z) = \left(\exp\left[-\frac{i}{2} x_n^{3/2}\left(\frac{1}{3}z^3 + z\right)\right], \exp\left[\frac{i}{2} x_n^{3/2}\left(\frac{1}{3}z^3 + z\right)\right]\right)\sigma_1 \exp(-\varphi\sigma_3), \tag{A2.34}$$

$$z \longrightarrow \pm\infty ,$$

where

$$\varphi = \frac{i}{2} x_n^{3/2}\left(-\frac{2\sqrt{2}}{3} + \frac{d_n}{4\sqrt{2}} + \frac{3d_n}{2\sqrt{2}}\ln 2 - \frac{d_n}{4\sqrt{2}}\ln d_n\right),$$

$$\alpha_n = 40 d_n x_n^{-2} + \frac{2}{9} \quad .$$

Thus the WKB-solutions are matched with canonical solutions as $x_n \to +\infty$

$$y_K(z) = y_{WKB_K}(z)\left[1 + O(x_n^{-2\delta})\right] C_K, \quad K = 1, 4, \tag{A2.35}$$

$$C_1 = \exp(\varphi\sigma_3)\cdot\begin{pmatrix} 1 & 0 \\ 0 & -i \end{pmatrix},$$

$$C_4 = \exp(\varphi\sigma_3)\sigma_1\begin{pmatrix} 1 & 0 \\ 0 & -i \end{pmatrix}.$$

In order to match the solutions (A2.34) y_{WKB_1} and y_{WKB_4} it is necessary to consider a neighbourhood of the origin $z = 0$. Omitting the small terms as $|z| \leq O(x_n^{-3/4+\delta})$, we can present the equation (A2.27) in this neighbourhood in the form

$$\frac{d^2 y_0}{dz^2} + \frac{1}{4} x_n^3 (2z^2 + \alpha_n) y_0 = 0 \,, \tag{A2.36}$$

which in its turn can be reduced to the Weber-Hermite equation

$$\frac{d^2 y_0}{d\zeta^2} + \left(\rho + \frac{1}{2} - \frac{\zeta^2}{4}\right) y_0 = 0 \,, \tag{A2.37}$$

where

$$z = e^{i\pi/4} x_n^{-3/4} 2^{-1/4} \zeta \,,$$

$$\rho + \frac{1}{2} = e^{i\pi/2} \frac{1}{4\sqrt{2}} x_n^{3/2} \alpha_n \,. \tag{A2.38}$$

The solutions of (A2.37) are expressed in terms of the Weber-Hermite functions

$$y_0(\zeta) = \left(D_\rho(\zeta),\; D_{-\rho-1}(i\zeta) \right) . \tag{A2.39}$$

This solution is related to an exact solution $y(z)$ of the equation (A2.27) through the following estimate, similar to those used above in (A2.18)

$$|y(z)-y_0(z)| \leq O(x_n^{-4\delta_1}) , \tag{A2.40}$$

as $|z|<O(x_n^{-9/16-\delta_1})$.

The matching domain for the solutions $y_0(z)$ and $y_{WKB}(z)$ is described by the relations

$$|z|=O(x_n^{-5/8}) , \ |\zeta|=O(x_n^{1/8}) , \ x_n \to +\infty ,$$

and hence we can replace the function $y_0(z)$ by its asymptotics at infinity

$$y_0(\zeta)=\left(e^{\zeta^2/4-(\rho+1)\ln\zeta}, e^{-\zeta^2/4+\rho\ln\zeta}\right), \left[1+O(\zeta^{-2})\right]M_k ,$$

$$M_1 = \sigma_1 \begin{pmatrix} 1 & 0 \\ 0 & e^{-i\pi/2(\rho+1)} \end{pmatrix} , \tag{A2.41}$$

$$M_4 = \sigma_1 \begin{pmatrix} 1 & , \dfrac{\sqrt{2\pi}}{\Gamma(\rho+1)} e^{-i\pi/2\rho} \\ \dfrac{\sqrt{2\pi}}{\Gamma(-\rho)} e^{i\pi(\rho+1)} & , e^{i3\pi/2(\rho+1)} \end{pmatrix} .$$

Applying again the calculation of asymptotics of the phase integral we have used above, one obtains the following expansions

$$y_{WKB_1}(\zeta)=A\left(e^{\zeta^2/4-(\rho+1)\ln\zeta}, e^{-\zeta^2/4+\rho\ln\zeta}\right)\left[1+O(\zeta^{-2})\right] \times$$
$$\times \exp\left[\sigma_3\left(-\psi-i\frac{\pi}{2}\left(\rho+\frac{1}{2}\right)\right)\right] , \tag{A2.42}$$
$$y_{WKB_4}(\zeta)= iA\left(e^{\zeta^2/4-(\rho+1)\ln\zeta}, e^{-\zeta^2/4+\rho\ln\zeta}\right)\left[1+O(\zeta^{-2})\right] \times$$
$$\times \exp\left[\sigma_3\left(-\psi+i\frac{\pi}{2}\left(\rho+\frac{1}{2}\right)\right)\right] ,$$

where

$$\psi=\frac{1}{2}(\rho+\frac{1}{2})-\frac{1}{2}(\rho+\frac{1}{2})\ln(\rho+\frac{1}{2}) ,$$

$$A=2^{7/8}x_n^{-1/8}(x-x_n)^{-1}e^{-i\pi/8} .$$

Therefore the matching matrices N_k , defined by the equations

$$y_{WKB_k}(\zeta)=Ay_0(\zeta)[1+O(\zeta^{-2})]N_k , \quad k=1,4 , \tag{A2.43}$$

are expressed by the following explicit formulae

$$N_1=\begin{pmatrix} 1 & 0 \\ 0 & e^{i\pi/2(\rho+1)}\end{pmatrix}\sigma_1 \exp[\sigma_3(-\psi-i\frac{\pi}{2}(\rho+\frac{1}{2}))] ,$$

(A2.44)

$$N_4=\begin{pmatrix} e^{2\pi i(\rho+1)} & , & -\frac{\sqrt{2\pi}}{\Gamma(\rho+1)}e^{i\pi/2} \\ -\frac{\sqrt{2\pi}}{\Gamma(-\rho)}e^{i3\pi/2(\rho+1)} & , & e^{i\pi/2(\rho+1)}\end{pmatrix}\sigma_1 \exp[\sigma_3(-\psi+i\frac{\pi}{2}(\rho+\frac{1}{2}))] .$$

Finally we have the leading term of asymptotics of the product $S_1S_2S_3$ as $x_n \to +\infty$ in the form

$$S_1S_2S_3=\begin{pmatrix} 1+\tau q & \tau \\ \tau & 1+\tau p\end{pmatrix}\simeq C_1^{-1}N_1^{-1}N_4C_4 =$$

$$=\begin{pmatrix} \frac{\sqrt{2\pi}}{\Gamma(-\rho)}e^{i\pi(\rho+1/2)+f} & , & e^{i\pi(\rho+1/2)} \\ e^{i\pi(\rho+1/2)} & , & \frac{\sqrt{2\pi}}{\Gamma(\rho+1)}e^{-f}\end{pmatrix} , \tag{A2.45}$$

$$f=i\frac{2\sqrt{2}}{3}x_n^{3/2}-\frac{3}{2}(\rho+\frac{1}{2})\ln x_n-\frac{7}{2}(\rho+\frac{1}{2})\ln 2-i\frac{\pi}{2}(\rho+\frac{1}{2}) .$$

Resolving the matrix equation (A2.45) with respect to r, p, q and applying again the formula for double argument of the Γ-function, we have

$$1+rp=\frac{1}{\sqrt{2}}\frac{\Gamma(\rho+\frac{1}{2})}{\Gamma(2\rho+1)}\exp\left[i\frac{\pi}{2}(\rho+\tfrac{1}{2})-h\right],$$

$$1+rq=\frac{1}{\sqrt{2}}\frac{\Gamma(-\rho-\frac{1}{2})}{\Gamma(-2\rho-1)}\exp\left[i\frac{\pi}{2}(\rho+\tfrac{1}{2})+h\right], \qquad \text{(A2.46)}$$

$$r=a_{-1}\cdot\exp[i\pi(\rho+\tfrac{1}{2})],\quad h=i\frac{2\sqrt{2}}{3}x_n^{3/2}-\frac{3}{2}(\rho+\tfrac{1}{2})\ln x_n-\frac{11}{2}(\rho+\tfrac{1}{2})\ln 2.$$

Just as above for the case of $x_n\to-\infty$ the isomonodromic condition implies the following constraints

$$\rho+\frac{1}{2}\in i\mathbb{R},\quad r=\bar{r},\quad \operatorname{sign} r=a_{-1},\quad r\neq 0, \qquad \text{(A2.47)}$$

$$p=\bar{q}, \qquad \text{(A2.47')}$$

which are equivalent to the real-valued reduction for the Ψ-function in (1.9). Moreover, the condition $r\neq 0$ means that the Stokes matrix S_2 is non-trivial.

Resolving the equations (A2.46) with respect to x_n, we obtain the asymptotic distribution of poles of the Painlevé function (A2.1) as $x_n\to+\infty$:

$$x_n^{3/2}=\frac{3}{\sqrt{2}}\pi n-\frac{3}{2\sqrt{2}}\beta\ln\frac{3}{\sqrt{2}}\pi n-\frac{21}{4\sqrt{2}}\beta\ln 2-$$

$$-\frac{3}{2\sqrt{2}}\arg\Gamma\left(\frac{1}{2}-i\beta\right)-\frac{3}{2\sqrt{2}}\chi,\quad n\to\infty, \qquad \text{(A2.48)}$$

where

$$\beta=\frac{1}{\pi}\ln|r|,\quad \chi=\arg(1+rp).$$

The formula (A2.47) shows that all the poles x_n have the same residues

$$a_{-1} = \operatorname{sign} \tau \,. \tag{A2.49}$$

REMARK. The assumption $0<|x-x_n|\leq O(e^{-|x_n|})$ may be weakened, replacing it by $0<|x-x_n|\leq O(|x_n|^{-1-\varepsilon})$, $\varepsilon>0$. All the asymptotic formulae obtained above remain true but their remainder terms would be altered.

Bringing together the results of this Appendix with those obtained in Chapters 4, 7 we can propose the following description of the singular solutions of the Painlevé II equation (A2.1):

THEOREM A2.1. Let $u(x)$ is the real-valued solution to the equation (A2.1) and p is the corresponding monodromy parameter. Then,

1. if $|p|<1$, $\operatorname{Re} p \neq 0$,

$$u(x)=\sqrt{2}\,\alpha(-x)^{-1/4}\sin\left\{\frac{2}{3}(-x)^{3/2}-\frac{3}{2}\alpha^2\ln(-x)+\varphi\right\}+o(x^{-1/2})$$

$$x\to-\infty$$

$$\alpha>0\,,\quad \alpha^2=-\frac{1}{2\pi}\ln(1-|p|^2)\,, \tag{A2.50}$$

$$\varphi=-3\alpha^2\ln 2-\frac{\pi}{4}-\arg\Gamma(-i\alpha^2)-\arg p$$

and

$$u(x)\approx\frac{a_{-1}}{x-x_n}\,,\quad x\sim x_n\,,\quad x_n\to+\infty\,,$$

$$x_n^{3/2}=\frac{3}{\sqrt{2}}\pi n-\frac{3}{2\sqrt{2}}\beta\ln\frac{3}{\sqrt{2}}\pi n-\frac{21}{4\sqrt{2}}\beta\ln 2-$$

$$-\frac{3}{2\sqrt{2}}\arg\Gamma\left(\frac{1}{2}-i\beta\right)-\frac{3}{2\sqrt{2}}\arg\frac{1+p^2}{1-|p|^2}+o(1)\,,\quad n\to\infty$$

$$\beta=\frac{1}{\pi}\ln\left|\frac{2\operatorname{Re}p}{1-|p|^2}\right|\,,\quad a_{-1}=\operatorname{sign}\operatorname{Re}p\,.$$

2. if $|p|>1$, $\operatorname{Re} p=0$,

$$u(x)\approx \frac{a_{-1}}{x-x_n}, \quad x\sim x_n, \quad x_n\to -\infty,$$

$$(-x_n)^{3/2}=3\pi n-\frac{3\alpha}{2}\ln 3\pi n-\frac{9}{2}\alpha\ln 2+\frac{3}{2}\arg\Gamma\left(\frac{1}{2}+i\alpha\right)-$$

$$-\frac{3\pi}{4}(1-a_{-1})+\frac{3}{2}\arg p+o(1), \quad n\to\infty$$

$$\alpha=\frac{1}{2\pi}\ln(|p|^2-1), \quad a_{-1}=(-1)^{n_0+n},$$

(A2.51)

n_0 – sufficiently large

and

$$u(x)=\frac{ip}{2\sqrt{\pi}}x^{-1/4}e^{-2/3x^{3/2}}\left(1+o(1)\right), \quad x\to+\infty .$$

3. if $|p|>1$, $\operatorname{Re} p\neq 0$ ("Scattering of poles"),

$$u(x)\approx \frac{a^-_{-1}}{x-x^-_n}, \quad x\sim x^-_n, \quad x^-_n\to -\infty,$$

$$(-x^-_n)^{3/2}=3\pi n-\frac{3\alpha}{2}\ln 3\pi n-\frac{9}{2}\alpha\ln 2+\frac{3}{2}\arg\Gamma\left(\frac{1}{2}+i\alpha\right)-$$

$$-\frac{3\pi}{4}(1-a_{-1})+\frac{3}{2}\arg p+o(1), \quad n\to\infty$$

(A2.52)

$$\alpha=\frac{1}{2\pi}\ln(|p|^2-1), \quad a^-_{-1}=(-1)^{n_0+n}$$

and

$$u(x)\approx \frac{a^+_{-1}}{x-x^+_n}, \quad x\sim x^+_n, \quad x^+_n\to +\infty,$$

$$x_n^{+\,3/2}=\frac{3}{\sqrt{2}}\pi n-\frac{3}{2\sqrt{2}}\beta\ln\frac{3}{\sqrt{2}}\pi n-\frac{21}{4\sqrt{2}}\beta\ln 2-\frac{3}{2\sqrt{2}}\arg\Gamma\left(\frac{1}{2}-i\beta\right)-$$

$$-\frac{3}{2\sqrt{2}}\arg\frac{1+p^2}{1-|p|^2}+o(1)\quad,\ n\to\infty\ ,$$

$$\beta=\frac{1}{\pi}\ln\left|\frac{2\operatorname{Re}p}{1-|p|^2}\right|\ ,\quad a^{+}_{-1}=-\operatorname{sign}\operatorname{Re}p\,.$$

Finally, we cencern a little the asymptotic behaviour of the solutions from the one-parameter submanifold described by the condition $|p|=1$. As it shown in [34] if $p=\mp i$, $r=0$, then

$$u(x)=\pm\left[\left(-\frac{x}{2}\right)^{1/2}-\frac{1}{2^{7/2}}(-x)^{-5/2}+O\left((-x)^{-11/2}\right)\right],\quad x\to-\infty$$

and (A2.53)

$$u(x)=\pm\frac{1}{2\sqrt{\pi}}x^{-1/4}e^{-2/3\,x^{3/2}}\left(1+o(1)\right)\ ,\ x\to+\infty\,.$$

If $p=\mp i$, $r\neq 0$, we propose a following hypothesis:

$$u(x)=\pm\left[\left(-\frac{x}{2}\right)^{1/2}-\frac{1}{2^{7/2}}(-x)^{-5/2}+O\left((-x)^{-11/2}\right)\right]\pm$$

$$\pm\frac{r}{2\sqrt{2\pi}}(-x)^{-1/4}\exp\left\{-\frac{2\sqrt{2}}{3}(-x)^{3/2}\right\}\left(1+o(1)\right)$$

$$x\to-\infty\ ,$$

and (A2.54)

$$u(x)\approx\frac{\pm a_{-1}}{(x-x_n)}\quad,\quad x\sim x_n\ ,\ x_n\to+\infty$$

$$x_n^{3/2}=\frac{3}{\sqrt{2}}\pi n-\frac{3}{2\sqrt{2}}\beta\ln\frac{3}{\sqrt{2}}\pi n-\frac{21}{4\sqrt{2}}\beta\ln 2-$$

$$-\frac{3}{2\sqrt{2}}\arg\Gamma\left(\frac{1}{2}-i\beta\right)-\frac{3}{2\sqrt{2}}\operatorname{arctg}r+o(1)\ ,\ n\to\infty\ ,$$

$$\beta=\frac{1}{\pi}\ln|r|\ ,\quad a_{-1}=\operatorname{sign}r\,.$$

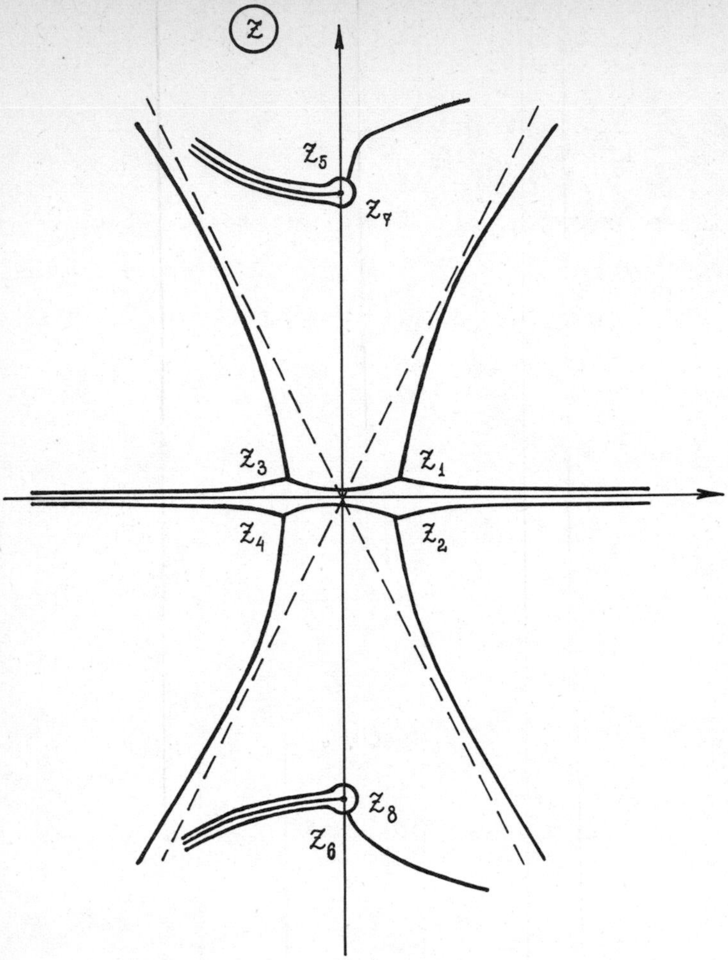

Figure A.1

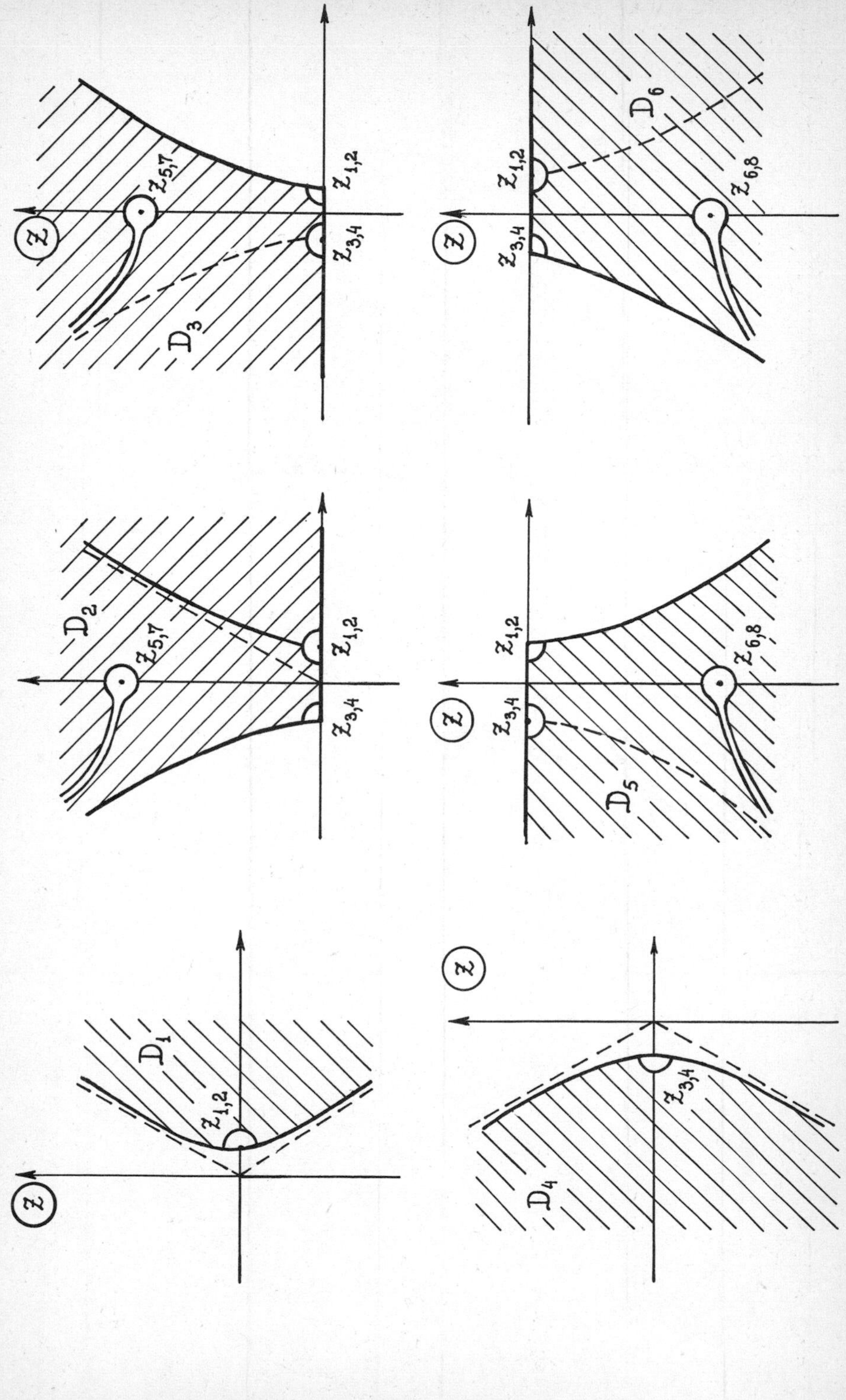

Figure A.2

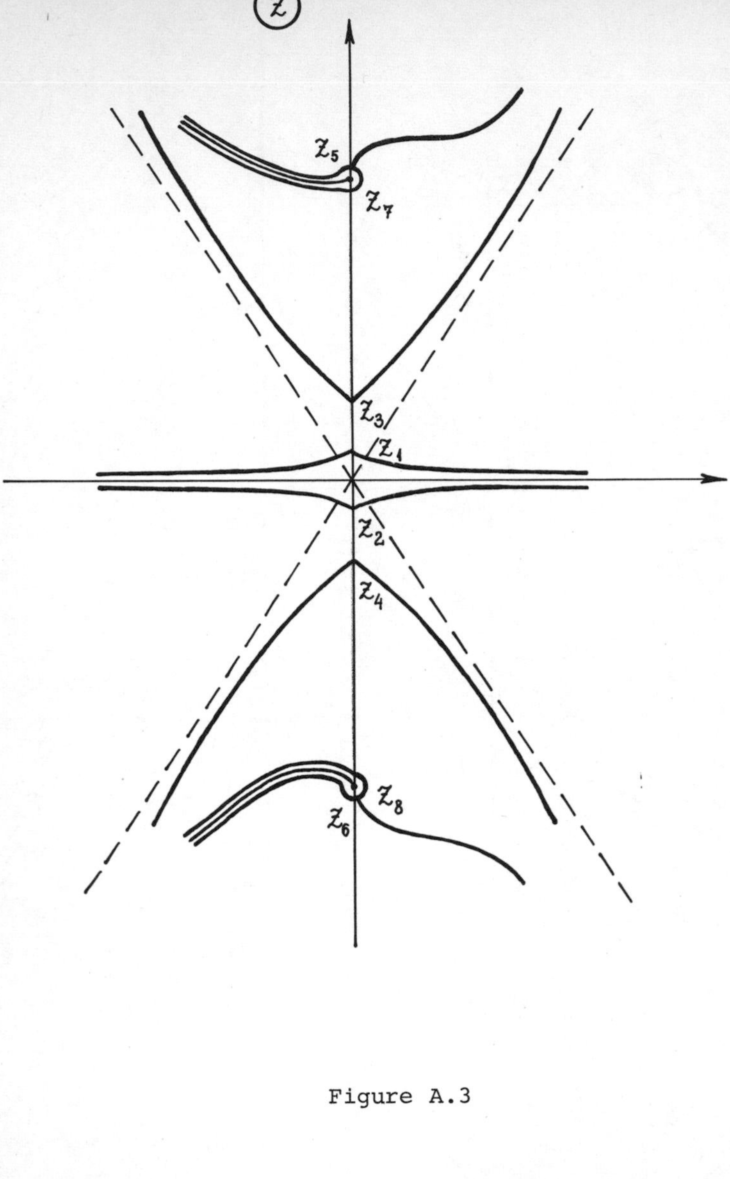

Figure A.3

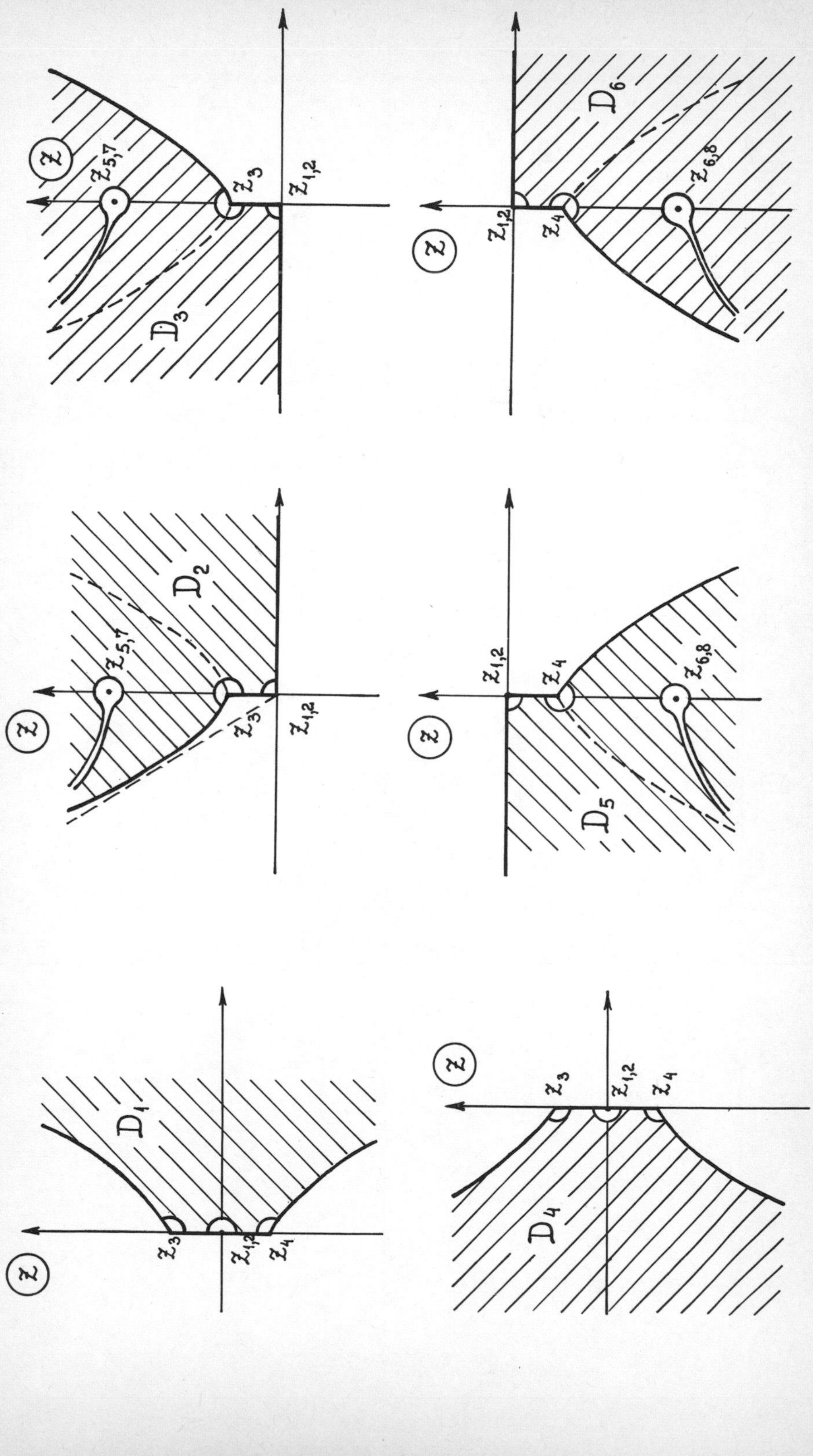

Figure A.4

APPENDIX 3. THE LIST OF PAINLEVÉ EQUATIONS AND THE CORRESPONDING "EQUATIONS IN λ "

Painlevé equation	A-matrix in the corresponding λ-equation $d\Psi/d\lambda = A(\lambda)\Psi$	Reference
PI : $u_{xx} = 6u^2 + x$	$A(\lambda) = \begin{pmatrix} 4 & 0 \\ 0 & -4 \end{pmatrix}\lambda^4 + \begin{pmatrix} 0 & -4u \\ 4u & 0 \end{pmatrix}\lambda^2 + \begin{pmatrix} 0 & -2u_x \\ -2u_x & 0 \end{pmatrix}\lambda + \begin{pmatrix} 1 & -1 \\ 1 & -1 \end{pmatrix}(2u^2 + x) - \frac{1}{2\lambda}\begin{pmatrix} 1 & -1 \\ 1 & -1 \end{pmatrix}$	[15]
PII : $u_{xx} = xu + 2u^3 + \nu$, $\nu \in \mathbb{C}$	$A(\lambda) = -4i\lambda^2\begin{pmatrix} 1 & 0 \\ 0 & -1 \end{pmatrix} + 4i\lambda\begin{pmatrix} 0 & u \\ -u & 0 \end{pmatrix} + \begin{pmatrix} -ix - 2iu^2 & -2u_x \\ -2u_x & ix + 2iu^2 \end{pmatrix} + \frac{1}{\lambda}\begin{pmatrix} 0 & -i\nu \\ i\nu & 0 \end{pmatrix}$	[16]

PIII : $u_{xx}=\frac{1}{u}u_x^2-\frac{1}{x}u_x+$ $+\frac{1}{x}(\alpha u^2+\beta)+4u^3-\frac{4}{u}$, $\alpha,\beta\in\mathbb{C}$	$A(\lambda)=\frac{x}{2}\begin{pmatrix}1 & 0\\ 0 & -1\end{pmatrix}+\frac{1}{\lambda}\begin{pmatrix}-\theta & y\\ v & \theta\end{pmatrix}+\frac{1}{\lambda^2}\begin{pmatrix}z-x/2 & -wz\\ \frac{z-x}{w} & -z+x/2\end{pmatrix}$ $u=-y/zw$, $\theta=(4-\beta)/8$ REMARK. $\frac{1}{8}\alpha\cdot\sigma_3=T_0^{\circ}$ – the exponent of formal monodromy at the zero.	[15]
The particular case of PIII : $(\alpha=\beta=0, u\to ie^{iu/2}, x\to x2^{-3/2})$ $u_{xx}+\frac{1}{x}u_x+\sin u=0$	$A(\lambda)=-i\frac{x^2}{16}\begin{pmatrix}1 & 0\\ 0 & -1\end{pmatrix}-\frac{1}{4\lambda}\begin{pmatrix}0 & ixu_x\\ ixu_x & 0\end{pmatrix}+$ $+\frac{1}{\lambda^2}\begin{pmatrix}i\cos u & -\sin u\\ \sin u & -i\cos u\end{pmatrix}$	[16]

PIV : $u_{xx} = \frac{1}{2u} u_x^2 + \frac{3}{2} u^3 + 4xu^2 + 2(x^2-\alpha)u + \beta/u$, $\alpha, \beta \in \mathbb{C}$.	$A(\lambda) = \lambda \begin{pmatrix} 1 & 0 \\ 0 & -1 \end{pmatrix} + \begin{pmatrix} x & y \\ \frac{2(z-\theta_0-\theta_\infty)}{y} & -x \end{pmatrix} + \frac{1}{\lambda} \begin{pmatrix} -z+\theta_0 & -u\cdot y/2 \\ \frac{2z(z-2\theta_0)}{u\cdot y} & z-\theta_0 \end{pmatrix}$ $\theta_\infty = (\alpha+1)/2, \quad \theta_0 = \sqrt{-\beta/8}$ $(T_0^\infty = \theta_\infty \sigma_3, \quad T_0^0 = \theta_0 \sigma_3)$	[15]
	$A(\lambda) = -8i\lambda^3 \begin{pmatrix} 1 & 0 \\ 0 & -1 \end{pmatrix} + 8\lambda^2 \begin{pmatrix} 0 & w \\ v & 0 \end{pmatrix} + i\lambda \begin{pmatrix} -2\sqrt{i}x - 4vw & 0 \\ 0 & 2\sqrt{i}x + 4vw \end{pmatrix} + \sqrt{i} \begin{pmatrix} 0 & 4xw + 2w_x \\ 4xv - 2v_x & 0 \end{pmatrix} + \frac{2}{\lambda} \begin{pmatrix} -i\theta_0 & 0 \\ 0 & i\theta_0 \end{pmatrix}$	[10]

	$u = v \cdot w/\sqrt{i}\ , \quad \theta_0 = (i\alpha + \beta/2)/2\ .$ $\left(T_0^{\infty} = -\left(\frac{3}{4} i\beta - \alpha/2\right)\sigma_3\ , \quad T_0^{0} = -2i\theta_0\sigma_3\right)$	
PV : $u_{xx} = \frac{3u-1}{2u(u-1)} u_x^2 - \frac{1}{x} u_x +$ $+ \frac{(u-1)^2}{x^2}(\alpha u + \beta/u) +$ $+ \frac{\gamma u}{x} + \frac{\delta u(u+1)}{u-1}\ ,$ $\alpha, \beta, \gamma \in \mathbb{C}\ , \quad \delta = -1/2$ - - general PV with non-zero δ is reduced to this case by scaling.	$A(\lambda) = \frac{x}{2}\begin{pmatrix} 1 & 0 \\ 0 & -1 \end{pmatrix} + \frac{1}{\lambda}\begin{pmatrix} z + \theta_0/2 & -v(z+\theta_0) \\ z/v & -z - \theta_0/2 \end{pmatrix} +$ $+ \frac{1}{\lambda - 1}\begin{pmatrix} -z - \frac{\theta_0 + \theta_\infty}{2} & vu\left(z + \frac{\theta_0 - \theta_1 + \theta_\infty}{2}\right) \\ -\frac{1}{vu}\left(z + \frac{\theta_0 + \theta_1 + \theta_\infty}{2}\right) & z + \frac{\theta_0 + \theta_\infty}{2} \end{pmatrix}$ $\alpha = \frac{1}{2}\left(\frac{\theta_0 - \theta_1 + \theta_\infty}{2}\right)^2\ , \quad \beta = -\frac{1}{2}\left(\frac{\theta_0 - \theta_1 - \theta_\infty}{2}\right)^2$ $\gamma = 1 - \theta_0 - \theta_1$ $\left(T_0^{\nu} = \frac{1}{2}\theta_\nu \sigma_3\ , \quad \nu = 0, 1, \infty\right)$	[15]

The particular case of PV :

$\alpha=\beta=0, \quad \gamma=2i,$

$\delta=2$

$$A(\lambda)=-i\tau\begin{pmatrix}1 & 0\\ 0 & -1\end{pmatrix}+\frac{1}{\lambda}\begin{pmatrix}1/4 & w\\ v & -1/4\end{pmatrix}+\frac{1}{\lambda^2}\begin{pmatrix}z & y\\ y & -z\end{pmatrix}$$

$$\tau=-\frac{x^2}{8}, \quad z=-\frac{i}{8}\frac{h+1}{h-1}, \quad y=-\frac{1}{4}\frac{\sqrt{h}}{h-1},$$

$$w+v=-\frac{i}{2\sqrt{h}}, \quad w-v=\frac{ih_\tau}{\sqrt{h}(1-h)}$$

$$h(\tau)=1-\frac{2}{2i\omega_x+\omega^2+1}, \quad \omega=\frac{\sqrt{u}+1}{\sqrt{u}-1}$$

REMARK. In this case the monodromy data are determined through the canonical solutions for which the calibration matrix B^∞ is not equal to I. It has the non-trivial dependence on $u(x)$ (see Appendix 1) :

$$B^\infty=\tau^{\sigma_3/8}\exp\left\{\frac{1}{8}\sigma_3\int^\tau\frac{dt}{h(t)t}\right\}.$$

[67]

PVI :		
$u_{xx}=\frac{1}{2}\left(\frac{1}{u}+\frac{1}{u-1}+\frac{1}{u-x}\right)u_x^2-$ $-\left(\frac{1}{x}+\frac{1}{x-1}+\frac{1}{u-x}\right)u_x+$ $+\frac{u(u-1)(u-x)}{x^2(x-1)^2}\left(\alpha+\beta\frac{x}{u^2}+\right.$ $\left.+\gamma\frac{x-1}{(u-1)^2}+\delta\frac{x(x-1)}{(u-x)^2}\right)$, $\alpha,\beta,\gamma,\delta\in\mathbb{C}$.	$A(\lambda)=\frac{1}{\lambda}\begin{pmatrix} z_0+\theta_0 & -yz_0 \\ y^{-1}(z_0+\theta_0) & -z_0\end{pmatrix}+\frac{1}{\lambda-1}\begin{pmatrix} z_1+\theta_1 & -vz_1 \\ v^{-1}(z_1+\theta_1) & -z_1\end{pmatrix}+$ $+\frac{1}{\lambda-x}\begin{pmatrix} z_2+\theta_2 & -wz_2 \\ w^{-1}(z_2+\theta_2) & -z_2\end{pmatrix}$, $z_0+z_1+z_2=k$ $yz_0+vz_1+wz_2=0$ $y^{-1}(z_0+\theta_0)+v^{-1}(z_1+\theta_1)+w^{-1}(z_2+\theta_2)=0$ $u=yz_0\cdot x/(x+1)yz_0+xvz_1+wz_2$ $k=-\frac{\theta_0+\theta_1+\theta_2+\theta_\infty}{2},\ \alpha=\frac{1}{2}(\theta_\infty-1)^2,\ \beta=\frac{1}{2}\theta_0^2,\ \gamma=\frac{1}{2}\theta_1^2,\ \delta=\frac{1}{2}(1-\theta_2^2)$ REMARK. This λ-equation posesses only regular singular points and the deformation parameter x coinsiding with a coordinate of one of the poles. Then (see the Introduction) we find the principled difficulties in treating the PVI equation with the IDM's technique.	[15]

APPENDIX 4. THE LIST OF CONNECTION FORMULAE FOR THE SOLUTIONS TO PAINLEVÉ EQUATIONS

PII equation : $u_{xx} = xu + 2u^3$

N	Asymptotics as $x \longrightarrow -\infty$	Asymptotics as $x \longrightarrow +\infty$	(n^o of formula) in the text, $[n^o]$ - reference
	Connection formulae		
1.	$u(x) = i\alpha(-x)^{-1/4}\sin\left\{\frac{2}{3}(-x)^{3/2} + \frac{3}{4}\alpha^2\ln(-x) + \varphi\right\} + o\left((-x)^{-1/4}\right)$,	$u(x) = i\varepsilon\sqrt{\frac{x}{2}} + i\varepsilon(2x)^{-1/4}\rho\cos\left\{\frac{2\sqrt{2}}{3}x^{3/2} - \frac{3}{2}\rho^2\ln x + \theta\right\} + o(x^{-1/4})$,	(9.32)
	$\alpha > 0$, $\varphi \neq \frac{3}{2}\alpha^2\ln 2 - \arg\Gamma\left(\frac{i\alpha^2}{2}\right) - \frac{\pi}{4}$, $\mathrm{mod}\,\pi$.	$\rho > 0$, $0 \leq \theta < 2\pi$, $\varepsilon = \pm 1$.	(9.33)

$$\rho^2 = \alpha^2 - \frac{1}{2\pi}\ln(e^{\pi\alpha^2} - 1) - \frac{1}{\pi}\ln 2\,|\sin\psi|,$$

$$\theta = \frac{7}{2}\rho^2\ln 2 + \arg\Gamma(i\rho^2) - \pi/4 - \arg\{1 + (e^{\pi\alpha^2} - 1)\exp 2i\psi\},$$

$$\mathrm{mod}\,2\pi, \quad \varepsilon = -\mathrm{sign}\,\sin\psi,$$

where

$$\psi = \frac{3}{2}\alpha^2\ln 2 - \arg\Gamma\left(\frac{i\alpha^2}{2}\right) - \pi/4 - \varphi.$$

No.	$x \to -\infty$	$x \to +\infty$	Reference
2.	$u(x)=i\alpha(-x)^{-1/4}\sin\{\frac{2}{3}(-x)^{3/2}+\frac{3}{4}\alpha^2\ln(-x)+\varphi\}+ o((-x)^{-1/4})$, $\alpha>0,\ \varphi=\frac{3}{2}\alpha^2\ln 2-\arg\Gamma(\frac{i\alpha^2}{2})-\frac{\pi}{4}+(1-\varepsilon)\frac{\pi}{2}$, $\mathrm{mod}\ 2\pi$, $\varepsilon=\pm 1$	$u(x)=\frac{ia}{2\sqrt{\pi}}x^{-1/4}e^{-2/3\,x^{3/2}}(1+o(1))$, $-\infty < a < \infty$.	(9.29) – (9.31)
	$a^2 = e^{\pi\alpha^2}-1$, $\mathrm{sign}\ a = \varepsilon$		
3.	$u(x)=\alpha(-x)^{-1/4}\cos\{\frac{2}{3}(-x)^{3/2}-\frac{3}{4}\alpha^2\ln(-x)+\varphi\}+ o((-x)^{-1/4})$, $\alpha>0,\ \varphi\neq-\frac{3}{2}\alpha^2\ln 2-\arg\Gamma(-\frac{i\alpha^2}{2})+\frac{\pi}{4}$, $\mathrm{mod}\ \pi$	$u(x)=\varepsilon_n\left[\frac{1}{x-x_n}-\frac{x_n}{6}(x-x_n)-\frac{1}{4}(x-x_n)^2+d_n(x-x_n)^3+O(x-x_n)^4\right]$, $x\sim x_n$, $\varepsilon_n=\pm 1$, $x_n\to+\infty$, $n\to\infty$	(7.9), (10.1), (10.4), (A2.48) (A2.50)
	$x_n^{3/2}=\frac{3}{\sqrt{2}}\pi n-\beta\ln n-\varkappa+o(1)$, $n\to\infty$; $d_n=\frac{1}{40}\left(\frac{2}{9}x_n^2+4\beta\sqrt{2x_n}\right)+o(1)$, $n\to\infty$; $\varepsilon_n=\mathrm{sign}\cos\Psi$, where		

	$\psi = -\frac{3}{2}\alpha^2 \ln 2 - \arg\Gamma\left(-\frac{i\alpha^2}{2}\right) - \frac{\pi}{4} - \varphi,$ $\beta = \frac{3}{2\sqrt{2}\pi} \ln\left\| \frac{\rho + \bar{\rho}}{1 - \|\rho\|^2} \right\|,$ $\varkappa = \beta \ln 24\pi + \frac{3}{2\sqrt{2}} \arg\Gamma\left(\frac{1}{2} - i\frac{2\sqrt{2}}{3}\beta\right) + \frac{3}{2\sqrt{2}} \arg \frac{1 + \rho^2}{1 - \|\rho\|^2}$ $\rho \equiv \left(1 - e^{-\pi\alpha^2}\right)^{1/2} e^{i\psi}$		
4.	$u(x) = \alpha(-x)^{-1/4} \cos\left\{\frac{2}{3}(-x)^{3/2} - \frac{3}{4}\alpha^2 \ln(-x) + \varphi\right\} +$ $+ o\left((-x)^{-1/4}\right),$ $\alpha > 0, \ \varphi = -\frac{3}{2}\alpha^2 \ln 2 - \arg\Gamma\left(-\frac{i\alpha^2}{2}\right) + \frac{\pi}{4} + (\varepsilon - 1)\frac{\pi}{2},$ $\mathrm{mod}\ 2\pi \qquad \varepsilon = \pm 1$	$u(x) = \frac{a}{2\sqrt{\pi}} x^{-1/4} e^{-\frac{2}{3}x^{3/2}}(1 + o(1)),$ $-1 < a < 1$	(4.9), (7.11) [31], [44]
	$a^2 = 1 - e^{-\pi\alpha^2}$ $\mathrm{sign}\, a = \varepsilon$		

5.	$u(x)=\varepsilon_n\left[\frac{1}{x-x_n}-\frac{x_n}{6}(x-x_n)-\frac{1}{4}(x-x_n)^2+ +d_n(x-x_n)^3+O(x-x_n)^4\right]$, $x \sim x_n$, $\varepsilon_n=\pm 1$, $x_n \to -\infty,\ n \to \infty$	$u(x)=\frac{a}{2\sqrt{\pi}}x^{-1/4}e^{-2/3 x^{3/2}}(1+o(1))$ $a \in \mathbb{R},\ \lvert a\rvert>1.$	(4.9), (A.2.26́), (A.2.26'), (A.2.51)

$$(-x_n)^{3/2}=3\pi n-\gamma \ln n-\theta+o(1)\ ,\quad n\to\infty$$

$$d_n=\frac{1}{40}\left(\frac{7}{9}x_n^2+\frac{16}{3}\gamma\sqrt{-x_n}\right)+o(1)\ ,\ n\to\infty$$

$$\varepsilon_n=(-1)^{n_0+n}\ ,\quad n_0\in\mathbb{N}\ ,$$

where

$$\gamma=\frac{3}{4\pi}\ln(a^2-1)\ ,$$

$$\theta=\gamma\ln 2\,4\pi-\frac{3}{2}\arg\Gamma\left(\frac{1}{2}+i\frac{2}{3}\gamma\right)+\frac{3}{4}\pi(1-\varepsilon_n)+\frac{3}{4}\pi\cdot\operatorname{sign} a$$

6.

$$u(x)=\varepsilon_n^-\left[\frac{1}{x-x_n^-}-\frac{x_n^-}{6}(x-x_n^-)-\frac{1}{4}(x-x_n^-)^2+\right.$$

$$\left.+d_n^-(x-x_n^-)^3+O(x-x_n^-)^4\right],$$

$x \sim x_n^-$

$\varepsilon_n^- = \pm 1$

$$(-x_n^-)^{3/2}=3\pi n-\gamma \ln n-\theta+o(1),\quad n\to\infty,$$

$$\theta \neq \gamma \ln 24\pi-\frac{3}{2}\arg\Gamma\left(\frac{1}{2}+i\frac{2}{3}\gamma\right)+$$

$$+\frac{3}{4}\pi(1-\varepsilon_n^-)\pm\frac{3}{4}\pi.$$

$$\left(d_n^-=\frac{1}{40}\left(\frac{7}{9}(x_n^-)^2+\frac{16}{3}\gamma\sqrt{-x_n^-}\right)+o(1),\quad n\to\infty\right)$$

$$u(x)=\varepsilon_n^+\left[\frac{1}{x-x_n^+}-\frac{x_n^+}{6}(x-x_n^+)-\right.$$

$$\left.-\frac{1}{4}(x-x_n^+)^2+d_n^+(x-x_n^+)^3+O(x-x_n^+)^4\right],$$

$x \sim x_n^+$

$\varepsilon_n^+ = \pm 1$

$$x_n^+ \longrightarrow +\infty,\quad n\to\infty$$

$$(x_n^+)^{3/2}=\frac{3}{\sqrt{2}}\pi n-\beta \ln n-\varkappa+o(1),\quad n\to\infty$$

$$d_n^+=\frac{1}{40}\left(\frac{2}{9}(x_n^+)^2+4\beta\sqrt{2x_n^+}\right)+o(1),\quad n\to\infty$$

$$\varepsilon_n^+ = -\operatorname{sign}\cos\Psi$$

where

(10.1), (10.4), (A2.48), (A2.26), (A2.26'), (A2.52)

	$\psi = \frac{2}{3}\gamma\, 24\pi - \arg\Gamma\left(\frac{1}{2} + i\frac{2}{3}\gamma\right) + \frac{1}{2}\pi(1-\varepsilon_n^-) - \frac{2}{3}\theta$ $\beta = \frac{3}{2\sqrt{2}\pi} \ln\left\lvert \frac{\rho+\bar{\rho}}{1-\lvert\rho\rvert^2} \right\rvert$, $\varkappa = \beta \ln 24\pi + \frac{3}{2\sqrt{2}} \arg\Gamma\left(\frac{1}{2} - i\frac{2\sqrt{2}}{3}\beta\right) + \frac{3}{2\sqrt{2}} \arg\frac{1+\rho^2}{1-\lvert\rho\rvert^2}$ $\rho \equiv (1+e^{4\pi/3\,\gamma})^{1/2} e^{i\psi}$ (note that to opposite the N 3 we have now $\lvert\rho\rvert > 1$)		
7.	$u(x) = \sqrt{-\frac{x}{2}} - \frac{1}{2^{7/2}}(-x)^{-5/2} + \cdots$	$u(x) = \frac{1}{2\sqrt{\pi}} x^{-1/4} e^{-2/3\, x^{3/2}} (1+o(1))$	(A2.53) [44], [34]
? 8.	$u(x) = \sqrt{-\frac{x}{2}} - \frac{1}{2^{7/2}}(-x)^{-5/2} + \ \ldots\ldots \ +$ $+ \frac{b}{2\sqrt{2\pi}}(-x)^{-1/4} \exp\left\{-\frac{2\sqrt{2}}{3}(-x)^{3/2}\right\} \cdot (1+o(1))$ $-\infty < b < \infty$	$u(x) = \varepsilon_n \left[\frac{1}{x-x_n} - \frac{x_n}{6}(x-x_n) - \frac{1}{4}(x-x_n)^2 + \right.$ $x \sim x_n \quad \left. + d_n(x-x_n)^3 + O(x-x_n)^4\right]$ $\varepsilon_n = \pm 1$ $x_n \longrightarrow +\infty, \quad n \longrightarrow \infty$	(A2.54)

$$x_n^{3/2} = \frac{3}{\sqrt{2}}\pi n - \beta \ln n - \varkappa + o(1), \quad n \to \infty$$

$$d_n = \frac{1}{40}\left(\frac{2}{9}(x_n)^2 + 4\beta\sqrt{2x_n}\right) + o(1), \quad n \to \infty$$

$$\varepsilon_n = \operatorname{sign} b \ , \qquad \beta = \frac{3}{2\sqrt{2}\pi} \ln|b| ,$$

$$\varkappa = \beta \ln 24\pi + \frac{3}{2\sqrt{2}} \arg\Gamma\left(\frac{1}{2} - i\frac{2\sqrt{2}}{3}\beta\right) - \frac{3}{2\sqrt{2}} \operatorname{arctg} b .$$

PIII equation : $u_{xx} + \frac{1}{x}u_x + \sin u = 0$

N	Asymptotics as $x \to 0$	Asymptotics as $x \to +\infty$	(N of formula) in the text, N - reference
	Connection formulae		
9.	$u(x) = r\ln x + s + O(x^2)$, $\operatorname{Im} r = \operatorname{Im} s = 0$	$u(x) = 2\pi k + \frac{d}{\sqrt{x}} \cos\left\{x - \frac{d^2}{16}\ln x + \varphi\right\} + o(x^{-1/2})$ $d > 0$, $0 \leqslant \varphi < 2\pi$	(8.35), (8.36), (8.37) [40]

$$\alpha^2 = -\frac{16}{\pi} \ln \frac{\mathrm{Re}A}{\pi} ,$$

$$\varphi = -\frac{\alpha^2}{8} \ln 2 - \arg\Gamma\left(-\frac{i\alpha^2}{16}\right) - \arg\rho + \frac{3\pi}{4} , \quad \mathrm{mod}\ 2\pi$$

$$\kappa = \mathrm{entier}\left\{\frac{1}{2\pi}\left[\pi - s + 3r\ln 2 - 4\arg\Gamma\left(\frac{1}{2} + \frac{ir}{4}\right)\right]\right\} ,$$

where

$$A = 2^{\frac{3ir}{2}} e^{\frac{is}{2}} \Gamma^2\left(\frac{1}{2} + \frac{ir}{4}\right), \quad \rho = \frac{Ae^{-\frac{\pi r}{4}} - Be^{\frac{\pi r}{4}}}{A+B} ,$$

$$B = 2^{-\frac{3ir}{2}} e^{-\frac{is}{2}} \Gamma^2\left(\frac{1}{2} - \frac{ir}{4}\right)$$

10.	$u(x) = r\ln x + s + O(x^{2-\lvert \mathrm{Im} r\rvert})$, $\mathrm{Re}\, r = \mathrm{Re}\, s = 0$, $\lvert \mathrm{Im} r\rvert < 2$.	$u(x) = \frac{i\alpha}{\sqrt{x}} \sin\left\{x + \frac{\alpha^2}{16}\ln x + \varphi\right\} + o(x^{-1/2})$, $\alpha > 0$	(8.5) , (8.11) , (8.37)

$$\alpha^2 = -\frac{8}{\pi} \ln\left(1 - \lvert\rho\rvert^2\right) ,$$

$$\varphi = +\frac{\alpha^2}{8} \ln 2 - \arg\Gamma\left(+\frac{i\alpha^2}{16}\right) - \arg\rho + \frac{3\pi}{4} , \quad \mathrm{mod}\ 2\pi$$

ρ – the same as in N 9.

11. $u(x) = r \ln x + s + O(x^{2-|\mathrm{Im}\, r|})$,

$|\mathrm{Im}\, r| < 2$, $\mathrm{Im}(A+B) = 0$,

$\arg \rho = -\arg q$,

where

$q = \dfrac{Be^{-\pi r/4} - Ae^{\pi r/4}}{A+B}$, and A, B, ρ –

the same as in N 9

$$u(x) = 2\pi k + \frac{\alpha}{2\sqrt{x}}\left[(\beta + \frac{1}{\beta})\cos\{x - \frac{\alpha^2}{16}\ln x + \varphi\} + i(\beta - \frac{1}{\beta})\sin\{x - \frac{\alpha^2}{16}\ln x + \varphi\}\right] + o(x^{-1/2})$$

$\alpha, \beta > 0$, $0 \leq \varphi < 2\pi$

(8.4), (8.10), (8.11), (8.37)

$$\alpha^2 = \frac{8}{\pi}\ln(1 + |\rho q|), \quad \beta^2 = \left|\frac{q}{\rho}\right|,$$

$$\varphi = -\frac{\alpha^2}{8}\ln 2 - \arg\Gamma(-\frac{i\alpha^2}{16}) - \arg\rho + \frac{3\pi}{4}, \quad \mathrm{mod}\ 2\pi$$

$$k = \mathrm{entier}\left\{\frac{1}{2\pi}\left[\pi - \mathrm{Re}\, s + 3\,\mathrm{Re}\, r \cdot \ln 2 - 4\arg\Gamma(\frac{1}{2} + \frac{ir}{4})\right]\right\}$$

(note that the solution from N 9 is the particular case ($\rho = \bar{q}$) of this solution)

12.	$u(x)=r\ln x+s+O(x^{2-\lvert \operatorname{Im} r\rvert})$, $\lvert \operatorname{Im} r\rvert<2$, $\operatorname{Im}(A+B)=0$, $\arg\rho=-\arg q+\pi$, $\boxed{A,B,\rho,q}$ – the same as in NN 9, 11	$u(x)=2\pi k+\frac{\alpha}{2\sqrt{x}}\Big[(\beta-\frac{1}{\beta})\cos\{x+\frac{\alpha^2}{16}\ln x+\varphi\}+$ $+i(\beta+\frac{1}{\beta})\sin\{x+\frac{\alpha^2}{16}\ln x+\varphi\}\Big]+o(x^{-1/2})$ $\alpha,\beta>0$, $0\leq\varphi<2\pi$	
	$\alpha^2=-\frac{8}{\pi}\ln(1-\lvert\rho q\rvert)$, $\beta^2=\left\lvert\frac{q}{\rho}\right\rvert$, $\varphi=\frac{\alpha^2}{8}\ln 2-\arg\Gamma(\frac{i\alpha^2}{16})-\arg\rho+\frac{3\pi}{4}$, mod 2π $k=\operatorname{entier}\Big\{\frac{1}{2\pi}\Big[\pi-\operatorname{Re}s+3\operatorname{Re}r\ln 2-4\arg\Gamma(\frac{1}{2}+i\frac{r}{4})\Big]\Big\}$ (note that the solution from N 10 is the particular case ($\rho=-\bar{q}$) of this solution)		
13.	$u(x)=r\ln x+s+O(x^{2-\lvert \operatorname{Im} r\rvert})$,	$u(x)=2\pi k+\frac{\alpha}{\sqrt{x}}\sqrt{\frac{2}{\pi}}\exp\{\mp ix\mp\frac{i\pi}{4}+i\varphi\}+o(x^{-1/2})$,	(8.6), (8.31),

	$\lvert \mathrm{Im}\, r\rvert < 2$, $\mathrm{Im}(A+B)=0$, $\exp is = 2^{-3ir}\Gamma^{2}\left(\frac{1}{2}-\frac{ir}{4}\right)\Gamma^{-2}\left(\frac{1}{2}+\frac{ir}{4}\right)\cdot$ $\cdot \exp\left(\mp\frac{\pi r}{2}\right)$	$\alpha>0$, $0\leqslant \varphi<2\pi$	(8.32) - - (8.33)
	$\alpha = 2\left\lvert \mathrm{sh}\frac{\pi r}{4}\right\rvert$, $\varphi = \frac{-1\pm 1}{2}\pi + \arg \mathrm{sh}\frac{\pi r}{4}$, $\mathrm{mod}\ 2\pi$, $K = \mathrm{entier}\left[\frac{1}{2\pi}\left\{\pi - \mathrm{Re}\,s + 3\,\mathrm{Re}\,r\cdot \ln 2 - 4\arg\Gamma\left(\frac{1}{2}+\frac{ir}{4}\right)\right\}\right]$ (note that this solution is the limiting case (q or $p\to 0$) of the solutions from NN 11, 12).		
14.	$u(x) = r\ln x + s + O(x^{2-\lvert \mathrm{Im}\, r\rvert})$, $\lvert \mathrm{Im}\, r\rvert < 2$, $\mathrm{Re}\, r = \mathrm{R}(s-\pi) = 0$, $e^{is} \neq -2^{-3ir}\Gamma^{2}\left(\frac{1}{2}-\frac{ir}{4}\right)\Gamma^{-2}\left(\frac{1}{2}+\frac{ir}{4}\right)$ $\Updownarrow$ $A+B = i\,\mathrm{Im}(A+B) \neq 0$	$u(x) = -2i\ln(x-x_n) + \pi - i\frac{x-x_n}{x_n} +$ $x\sim x_n$ $+ib_n\frac{(x-x_n)^2}{x_n^2} + O(x-x_n)^3$, $x_n\to\infty$, $n\to\infty$	(11.2), (11.6), (11.61), (11.66) [4I]

$$x_n = 2\pi n - \beta \ln n - \varkappa + o(1)\ , \quad n \to \infty\ ,$$

$$b_n = \frac{1}{6} x_n^2 + \frac{2}{3} x_n \beta - \frac{1}{2} + o(1)\ , \quad n \to \infty\ ,$$

where

$$\beta = \frac{1}{\pi} \ln \left| \frac{A+B}{2\pi} \right|\ ,$$

$$\varkappa = \beta \ln 8\pi - \arg \Gamma\left(\tfrac{1}{2} + i\beta\right) - \arg \rho\ ,$$

ρ – the same as in N 9.

15.

$$u(x) = r \ln x + s + O\left(x^{2-|\mathrm{Im}\, r|}\right)\ ,$$

$$|\mathrm{Im}\, r| < 2\ , \quad \mathrm{Re}\, r = \mathrm{Re}(s - \pi) = 0\ ,$$

$$e^{is} = -2^{-3ir}\, \Gamma^2\left(\tfrac{1}{2} - \tfrac{ir}{4}\right) \Gamma^{-2}\left(\tfrac{1}{2} + \tfrac{ir}{4}\right)$$

$$\Updownarrow$$

$$A + B = i\,\mathrm{Im}(A+B) = 0$$

$$u(x) = \pi + \frac{i\alpha}{\sqrt{x}} \sqrt{\frac{2}{\pi}}\, e^{-x} \left(1 + o(1)\right)\ ,$$

$$-2 < \alpha < 2$$

(11.3), (11.11), (11.12)

[5]

$$\alpha = 2 \sin\left(i \frac{\pi r}{4}\right)$$

16.	$u(x) = 2\varepsilon i \ln\left[-2x(\ln x + \gamma)\right] + \pi + O(x^4 \ln^2 x)$, $\varepsilon = \pm 1$, $\gamma = 0{,}577\ldots$ – Euler constant	$u(x) = \pi + i\frac{\alpha}{\sqrt{x}}\sqrt{\frac{2}{\pi}}\, e^{-x}(1 + o(1))$, $\lvert\alpha\rvert = 2$	[5]
	$\alpha = 2\varepsilon$		
17.	$\exp\left(-\frac{u(x)}{2}\right) = \sin\{2\mu[\ln x + \varphi]\} + o(1)$, $0 \neq \mu \in \mathbb{R}$, $\varphi = \frac{1}{\mu}\arg\Gamma(i\mu)$	$u(x) = \pi + \frac{i\alpha}{\sqrt{x}}\sqrt{\frac{2}{\pi}}\, e^{-x}(1 + o(1))$, $\alpha > 2$	[5]
	$\alpha = 2\operatorname{ch}\pi\mu$		

PIV equation : $$u_{xx}=\frac{1}{2}\frac{u_x^2}{u}-\frac{3}{8}u^3-\frac{x}{2}u^2-\left(\frac{x^2}{8}-\frac{\mu}{2}\right)u+\frac{\nu^2}{2u}$$

Asymptotics as $|x|\longrightarrow\infty$, $\arg x=\frac{\pi}{2}\ell$, $\ell=-2,0,-1,1$:

$$u(x)=x^{-1}\left(a_\ell+b_\ell e^{i\theta_\ell(x)}+c_\ell e^{-i\theta_\ell(x)}\right),$$

$$\theta_\ell(x)=\frac{x^2}{4}-4\left(\gamma+(-1)^{\ell+1}\frac{3}{2}\rho_\ell\right)\ln|x|+\left(2\gamma+(-1)^{\ell+1}\rho_\ell\right)\ln 2,$$

$$a_\ell=4\rho_\ell+2\beta-2\gamma,$$

$$b_\ell=2i\frac{(2\pi)^{3/2}\exp\{\pi(\frac{3}{2}\rho_\ell+(-1)^{\ell+1}\gamma)\}}{\rho_\ell\Gamma(i(-1)^\ell\rho_\ell)\cdot\Gamma(\frac{1}{2}-i\beta-i\gamma+i(-1)^\ell\rho_\ell)\cdot\Gamma(i\beta-i\gamma+i(-1)^\ell\rho_\ell)}$$

$$c_\ell=-2i\frac{(2\pi)^{3/2}\exp\{\pi(\frac{3}{2}\rho_\ell+(-1)^{\ell+1}\gamma)\}}{q_\ell\Gamma(-i(-1)^\ell\rho_\ell)\Gamma(\frac{1}{2}+i\beta+i\gamma-i(-1)^\ell\rho_\ell)\cdot\Gamma(i\gamma-i\beta-i(-1)^\ell\rho_\ell)}$$

where

$$(\gamma-\beta)^2=\nu^2 \quad , \quad \gamma+3\beta=\mu \quad , \quad \rho_\ell=\frac{1}{2\pi}\ln\alpha_\ell \quad ,$$

$$p_0=p_1=\alpha_0\alpha_1-1 \quad , \quad q_0=q_{-1}=\alpha_0\alpha_{-1}-1 \quad ,$$

$$p_{-2}=p_{-1}=\alpha_{-2}\alpha_{-1}-1 \, , \quad q_{-2}=q_1=\alpha_{-2}\alpha_1-1 \, ,$$

and α_ℓ , $\ell=-2,-1,0,1$ are asymptotic parameters

Connection formulae :

$$1-\alpha_0-\alpha_{-2} = 2\mathrm{Sh}\,2\pi\beta\, e^{2\pi\gamma}-\alpha_1\alpha_{-1}e^{4\pi\gamma}$$

$$1-\alpha_{-1}-\alpha_1 = -2\mathrm{Sh}\,2\pi\beta\, e^{-2\pi\gamma}-\alpha_0\alpha_{-2}e^{-4\pi\gamma}$$

Reference : [70]

PV equation : $u_{xx}=(u_x)^2\frac{3u-1}{2u(u-1)}-\frac{1}{x}u_x+\frac{2i}{x}u+2u\frac{u+1}{u-1}$

N°	Asymptotics as $x \to 0$	Asymptotics as $x \to +\infty$	(n° of formula) in the text, [n°]- reference
	Connection formulae		
18.	$u=\exp(2i\Phi)$		(A.1.5) - -(A.1.7), (A.1.62)- -(A.1.65)
	$\Phi=x-ax^2+O(x^3)$ $a<\frac{1}{\pi}$	$\Phi=x+k\ln x+x_0+o(1)$	
	$k=\frac{1}{\pi}\ln(1-a\pi)$, $x_0=-2\arg\Gamma\left(\frac{ik}{2}\right)+k\ln 2-\pi\,\mathrm{sign}\,k$, $k\neq 0$ $x_0=0$, $k=0$		
19.	$\Phi=x-ax^2+O(x^3)$, $a=\frac{1}{\pi}$	$\Phi=\frac{\pi}{2}+o(1)$	(A.1.7)
20.	$\Phi=x-ax^2+O(x^3)$, $a>\frac{1}{\pi}$	$\Phi=-x+k\ln x+x_0+o(1)$	(A.1.8), (A.1.79) - - (A.1.81)

$$k = -\frac{1}{\pi} \ln(\pi a - 1) \quad ,$$

$$x_0 = \frac{\pi}{2} + 2 \arg \Gamma\left(\frac{ik}{2} - \frac{1}{2} \right) + k \ln 2 + l\pi \; , \qquad l \in \mathbb{Z} \qquad [7]$$

REFERENCES

1. Painlevé P. Sur les équations différentielles du second ordre et d'ordre supérieur, dont l'intérable génerale est uniforme.- Acta Math., 1902, v.25, p.1-86.

2. Gambier B. Sur les équations différentielles du second ordre et du premier degré dont l'intégrale est á points critiques fixes.- Acta Math., 1910, v.33, p. 1-55.

3. Erugin N.P. The theory of the movable singularities of the second order equations.- Differ.Uravn., 1976, v.12, p.579-598.

4. Fokas A.S., Ablowitz M.J. On a unified approach to transformations and elementary solutions of Painlevé equations.- J.Math.Phys., 1982, v.23, N 11, p.2033-2042.

5. McCoy B.M., Tracy C.A., Wu T.T. Painlevé functions of the third kind.- J.Math.Phys., 1977, v.18, N 5, p.1058-1092.

6. McCoy B.M., Wu T.T. The two-dimensional Ising model.- Nucl.Phys., 1981, v.B180 (FS2), p.89.

7. Creamer D.B. Thacker H.B., Wilkinson D. Some exact results for the two-point functions of an integrable quantum field theory.- Phys. Rev.D, 1981, v.23, N 12, p.3081-3084.

8. Barouch E., McCoy B.M., Wu T.T. Zero-field susceptibility of the two-dimensional Ising model near T_c .- Phys.Rev.Lett., 1973, v.31, p.1409.

9. Perk J.H.H., Capel H.W., Quispel G.R.W., Nijhoff F.W. Finite-temperature correlations for the Ising chain in a transverse field.- Preprint ITP-SB-82-72 (Univ. of N.Y., Stony Brook).

10. Jimbo M., Miwa T., Mori Y., Sato M. The unanticipated link between deformation theory of differential equations and quantum fields.- Preprint RIMS-305, Kyoto, 1979.

11. Gromak V.I., Tsegelnik V.V. Nonlinear two-dimensional theoretical field's models and Painlevé equations.- Teor.Mat.Fiz., 1983, v.55, N 2, p.189-196.

12. Ablowitz M.J., Ramani A., Segur H. A connection between nonlinear evolution equations and ordinary differential equations of P-type I.- J.Math.Phys., 1980, v.21, N 4, p.715-721.

13. Ablowitz M.J., Ramani A., Segur H. A connection between nonlinear

evolution equations and ordinary differential equations of P-type II .- J.Math.Phys., 1980, v.21, N 5, p.1006-1015.

14. Jimbo M., Miwa T., Ueno K., Monodromy preserving deformation of linear ordinary differential equations with rational coefficients. I.- Preprint RIMS-319, Kyoto, 1980.

15. Jimbo M., Miwa T., Ueno K. Monodromy preserving deformation of linear ordinary differential equations with rational coefficients. II .- Preprint RIMS-327, Kyoto, 1980.

16. Flaschka H., Newell A.C. Monodromy and spectrum preserving deformations I.- Commun.Math.Phys., 1980, v.76, p.67-116.

17. Manakov S.V. On propagation of the pulse in the long laser amplifier - JETP Lett., 1982, v.35, N 5, p.103-195.

18. Manakov S.V., Gabitov I.R. Propagation of ultrashort optical pulses in degenerate laser amplifiers.- Phys.Rev.Lett., 1983, v.50, N 7, p.495-498.

19. Zakharov V.E., Kuznetsov E.A., Musher S.L. On the quasi-classical regime of the three dimensional wave collapse.- IETP Lett., 1985, v.41, N 3, p.125-127.

20. Tajiri M. On reductions to the second Painlevé equation and N-solution solutions of the two and three dimensional nonlinear Klein--Gordon equations.- J.Phys.Soc.Japan, 1984, v.53, N 1, p.1-4.

21. Fuchs R., - Math.Ann., 1907, v.63, p.301-321.

22. Garnier R., - Ann.Sci.Ec.Norm.Super., 1912, v.29, p.1-126.

23. Riemann B., Gesammelte Mathematische Werke. Leipzig, 1892.

24. Poincaré H. Sur les groupes des équations linéaires.- Acta Math., 1884, v.4, p.201-311.

25. Birkhoff G.D. A simplified treatment of the regular singular point. - Trans.Am.Math.Soc., 1910, p.199-202.

26. Lappo-Danilevskij I.A. An application of the functions on matrices to the theory of systems of linear ordinary differential equations. - Moscow, Acad.Sci. USSR, 1957.

27. Novokshenov V.Yu. The isomonodromic deformation method and asymptotics of the third Painlevé transcendent.- Funkts.Anal. Prolozh., 1984, v.18, N 3, p.90-91.

28. Its A.R., Petrov V.E. "Isomonodromic" solutions of the Sine-Gordon equation and the time asymptotics of its rapidly decreasing solu-

tions.- Sov.Math.Dokl., 1982, v.26, N 1, p.244-247.

29. Its A.R. Asymptotics of solutions of the nonlinear Schrödinger equation and isomonodromic deformations of systems of linear differential equations.- Sov.Math.Dokl., 1981, v.24, N 3, p.452-456.

30. Its A.R. "Isomonodromic" solutions of the zero-curvature's equations.- Izv.Akad.Nauk, SSSR, Ser.Mat., 1985, v.49, N 3, p.32-63.

31. Ablowitz M.J., Segur H. Exact linearization of a Painlevé transcendent.- Phys.Rev.Lett., 1977, v.38, p.1103-1106.

32. Novokshenov V.Yu., Suleimanov B.I. The isomonodromic deformation method and asymptotics of the second and third Painlevé transcendents.- Usp.Mat.Nauk, 1984, v.39, N 4, p.114-115.

33. Abdullaev A.S. On the theory of the second Painlevé equation.- Dokl.Akad. Nauk SSSR, 1983, v.273, N 5, p. 31-51.

34. Hastings S.P., McLeod J.B. A boundary value problem associated with the second Painlevé transcendent and the Korteveg-de Vries equation.- Arch.Ration.Mech.Anal., 1980, v.73, N 1, p.31-51.

35. Lukashevich N.A. On the theory of the third Painlevé equation.- Differ.Uravn., 1967, v.3, N 11, p.1913-1923.

36. Kimura H., Okamoto K. On the polynomial Hamiltonian structure of the Garnier systems.- J.Math.Pures Appl., 1984, v.63, N 1, p.129-146.

37. Fokas A.S., Ablowitz M.J. On the initial value problem of the second Painlevé transcendent.- Commun.Math.Phys., 1983, v.91, p.381-403.

38. Fedoriuk M.V. Asymptotic methods for linear ordinary differential equations.- Moscow: "Nauka" pub. 1983.

39. Its A.R., Kapaev A.A. The isomonodromic deformation method and connection formulae for the second Painlevé transcendent.- Izv. Akad.Nauk SSSR, Ser.Mat., 1986 (to be published).

40. Novokshenov V.Yu. On the asymptotics of the general real-valued solution to the third Painlevé equation.- Dokl.Akad.Nauk SSSR, 1985, v.283, N 5, p.1161-1165.

41. Novokshenov V.Yu. On the connection formulae for the solutions to the Painlevé equation of the third kind.- Funkts. Annal.Prilozh. 1986 (to be published).

42. Kapaev A.A., Novokshenov V.Yu. On the connection formulae for the

general real-valued solution to the second Painlevé equation.- Dokl.Akad.Nauk SSSR, 1986 (to be published).

43. Ablowitz M.J., Segur H. Asymptotic solutions of the Korteweg-de Vries equation.- Stud.Appl.Math., 1977, v.57, N 1, p.13-44.

44. Ablowitz M.J., Segur H. A note a Miura's transformation.- Preprint

45. Zakharov V.E., Manakov S.V. On the asymptotic behaviour of the nonlinear wave systems solvable by the inverse scattering method.- Z.Èksper.Teoret.Fiz., 1976, v.71, p.203-215.

46. Shabat A.B. On the Korteweg-de Vries equation.- Sov.Math.Dokl., 1973, v.14, p.1266.

47. Coddington E.A., Levinson N. Theory of ordinary differential equations.- McGraw-Hill, 1955.

48. Wasow W. Asymptotic expansions for ordinary differential equations. - N.Y.Interscience Pub., 1965.

49. Sibuya Y. Stokes phenomena.- Bull-Am.Math.Soc., 1977, v.83, p.1075-1077.

50. Schlesinger L. Über eine Klasse von Differentialsystemen beliebiger Ordnung mit festen kritischen Punkten.- J.Reine Angew.Math., 1912, v.141, p.96-145.

51. Zakharov V.E., Shabat A.B. Integration of the nonlinear equations by means of the inverse scattering method.- Funkts.Anal.Prilozh., 1979, v.13, N 3, p. 13-22.

52. Olver F.W.J. Asymptotics and special functions.- Academic Press N.Y., 1974.

53. Bateman H., Erdélyi A. Higher transcendental functions, V.2.- McGraw-Hill, N.Y., 1953.

54. Bender C.M., Wu T.T.Anharmonic oscillator.- Phys.Rev., second ser., 1969, v.184, N 5, p.1231.

55. Hinton F.L.- J.Math.Phys., 1979, v.20, N 10, p.2036-2046.

56. Bateman H., Erdélyi A. Higher transcendental functions, V.3.- Mcgraw-Hill, N.Y., 1953.

57. Bordag L.A. Painlevé equations and their connection with nonlinear evolution equations.- Preprint JINR, E5-80-477, 1980.

58. Lamb. G.L.- Jr.Rev.Mod.Phys., 1971, v.43, p.99.

59. Akhmanov S.A., Sukhorukov A.P., Khokhlov R.V. Self-focussing and self-trapping of intense light beams in a nonlinear medium.- Z. Èksper.Teoret.Fiz., 1966, v.50, p.1537-1549.

60. Zakharov V.E.- Z.Èksper.Teoret.Fiz., 1972, v.62, p.1745-1759.

61. Zhiber A.V., Zakharov V.E. On the initial value problem of the nonlinear Schrödinger equation.- Differ. Uravn., 1970, v.6, N 1, p.137-146.

62. Kamke E. Differentialgleichungen , lösungs-metoden und lösungen, v.1. Leipzig, 1959.

63. Manakov S.V. Nonlinear Fraunhoffer diffraction. - Z.Èksper.Teoret.Fiz., 1973, v.65, p.1392-1403.

64. Vaidya H.G., Tracy C.A. One particle reduced density matrix of impenetrable bosons in one dimension at zero temperature.- J.Math. Phys., 1979, v.20, N 11, p.2291-2312.

65. Gromak V.I. On the theory of the Painlevé equations.- Differ.Uravn. 1975, v.11, N 2, p.373-376.

66. Popowicz Z.- Lect.Notes Phys., 1983, v.189, p.416-423.

67. Airault H. An inverse problem for fifth Painlevé equation.- Comptes Rendus, 1982, v.294, p.185-188.

68. Okamoto K. Polynomial Hamiltoniana associated which Painlevé equations. I.- Proc.Jap.Acad., Ser.A, 1980, v.56, N 6, p.264-268.

69. Demidovich B.P. Lectures on the mathematical theory of elasticity. Moscow: "Nauka" pub., 1976.

70. Kitaev A.V. On the self-similar solutions of the modified non-linear Schrödinger equation. - Teor.Mat.Fiz., 1985, v.64, N 3, p.347-369.

71. Gakhov F.D. Boundary value problems. - New York: Pergamon Press 1966.

SUBJECT INDEX

Vol. 1090: Differential Geometry of Submanifolds. Proceedings, 1984. Edited by K. Kenmotsu. VI, 132 pages. 1984.

Vol. 1091: Multifunctions and Integrands. Proceedings, 1983. Edited by G. Salinetti. V, 234 pages. 1984.

Vol. 1092: Complete Intersections. Seminar, 1983. Edited by S. Greco and R. Strano. VII, 299 pages. 1984.

Vol. 1093: A. Prestel, Lectures on Formally Real Fields. XI, 125 pages. 1984.

Vol. 1094: Analyse Complexe. Proceedings, 1983. Edité par E. Amar, R. Gay et Nguyen Thanh Van. IX, 184 pages. 1984.

Vol. 1095: Stochastic Analysis and Applications. Proceedings, 1983. Edited by A. Truman and D. Williams. V, 199 pages. 1984.

Vol. 1096: Théorie du Potentiel. Proceedings, 1983. Edité par G. Mokobodzki et D. Pinchon. IX, 601 pages. 1984.

Vol. 1097: R.M. Dudley, H. Kunita, F. Ledrappier, École d'Éte de Probabilités de Saint-Flour XII – 1982. Edité par P.L. Hennequin. X, 396 pages. 1984.

Vol. 1098: Groups – Korea 1983. Proceedings. Edited by A.C. Kim and B.H. Neumann. VII, 183 pages. 1984.

Vol. 1099: C.M. Ringel, Tame Algebras and Integral Quadratic Forms. XIII, 376 pages. 1984.

Vol. 1100: V. Ivrii, Precise Spectral Asymptotics for Elliptic Operators Acting in Fiberings over Manifolds with Boundary. V, 237 pages. 1984.

Vol. 1101: V. Cossart, J. Giraud, U. Orbanz, Resolution of Surface Singularities. Seminar. VII, 132 pages. 1984.

Vol. 1102: A. Verona, Stratified Mappings – Structure and Triangulability. IX, 160 pages. 1984.

Vol. 1103: Models and Sets. Proceedings, Logic Colloquium, 1983, Part I. Edited by G.H. Müller and M.M. Richter. VIII, 484 pages. 1984.

Vol. 1104: Computation and Proof Theory. Proceedings, Logic Colloquium, 1983, Part II. Edited by M.M. Richter, E. Börger, W. Oberschelp, B. Schinzel and W. Thomas. VIII, 475 pages. 1984.

Vol. 1105: Rational Approximation and Interpolation. Proceedings, 1983. Edited by P.R. Graves-Morris, E.B. Saff and R.S. Varga. XII, 528 pages. 1984.

Vol. 1106: C.T. Chong, Techniques of Admissible Recursion Theory. IX, 214 pages. 1984.

Vol. 1107: Nonlinear Analysis and Optimization. Proceedings, 1982. Edited by C. Vinti. V, 224 pages. 1984.

Vol. 1108: Global Analysis – Studies and Applications I. Edited by Yu.G. Borisovich and Yu.E. Gliklikh. V, 301 pages. 1984.

Vol. 1109: Stochastic Aspects of Classical and Quantum Systems. Proceedings, 1983. Edited by S. Albeverio, P. Combe and M. Sirugue-Collin. IX, 227 pages. 1985.

Vol. 1110: R. Jajte, Strong Limit Theorems in Non-Commutative Probability. VI, 152 pages. 1985.

Vol. 1111: Arbeitstagung Bonn 1984. Proceedings. Edited by F. Hirzebruch, J. Schwermer and S. Suter. V, 481 pages. 1985.

Vol. 1112: Products of Conjugacy Classes in Groups. Edited by Z. Arad and M. Herzog. V, 244 pages. 1985.

Vol. 1113: P. Antosik, C. Swartz, Matrix Methods in Analysis. IV, 114 pages. 1985.

Vol. 1114: Zahlentheoretische Analysis. Seminar. Herausgegeben von E. Hlawka. V, 157 Seiten. 1985.

Vol. 1115: J. Moulin Ollagnier, Ergodic Theory and Statistical Mechanics. VI, 147 pages. 1985.

Vol. 1116: S. Stolz, Hochzusammenhängende Mannigfaltigkeiten und ihre Ränder. XXIII, 134 Seiten. 1985.

Vol. 1117: D.J. Aldous, J.A. Ibragimov, J. Jacod, Ecole d'Été Probabilités de Saint-Flour XIII – 1983. Édité par P.L. Henneq IX, 409 pages. 1985.

Vol. 1118: Grossissements de filtrations: exemples et applicati Seminaire, 1982/83. Edité par Th. Jeulin et M. Yor. V, 315 pages. 19

Vol. 1119: Recent Mathematical Methods in Dynamic Programm Proceedings, 1984. Edited by I. Capuzzo Dolcetta, W.H. Flem and T. Zolezzi. VI, 202 pages. 1985.

Vol. 1120: K. Jarosz, Perturbations of Banach Algebras. V, 118 pag 1985.

Vol. 1121: Singularities and Constructive Methods for Their Treatm Proceedings, 1983. Edited by P. Grisvard, W. Wendland and Whiteman. IX, 346 pages. 1985.

Vol. 1122: Number Theory. Proceedings, 1984. Edited by K. All VII, 217 pages. 1985.

Vol. 1123: Séminaire de Probabilités XIX 1983/84. Proceedings. E par J. Azéma et M. Yor. IV, 504 pages. 1985.

Vol. 1124: Algebraic Geometry, Sitges (Barcelona) 1983. Proc ings. Edited by E. Casas-Alvero, G.E. Welters and S. Xam Descamps. XI, 416 pages. 1985.

Vol. 1125: Dynamical Systems and Bifurcations. Proceedings, 19 Edited by B.L.J. Braaksma, H.W. Broer and F. Takens. V, 129 pag 1985.

Vol. 1126: Algebraic and Geometric Topology. Proceedings, 19 Edited by A. Ranicki, N. Levitt and F. Quinn. V, 523 pages. 1985.

Vol. 1127: Numerical Methods in Fluid Dynamics. Seminar. Editec F. Brezzi, VII, 333 pages. 1985.

Vol. 1128: J. Elschner, Singular Ordinary Differential Operators Pseudodifferential Equations. 200 pages. 1985.

Vol. 1129: Numerical Analysis, Lancaster 1984. Proceedings. Ed by P.R. Turner. XIV, 179 pages. 1985.

Vol. 1130: Methods in Mathematical Logic. Proceedings, 1983. Ed by C.A. Di Prisco. VII, 407 pages. 1985.

Vol. 1131: K. Sundaresan, S. Swaminathan, Geometry and Nonlir Analysis in Banach Spaces. III, 116 pages. 1985.

Vol. 1132: Operator Algebras and their Connections with Topol and Ergodic Theory. Proceedings, 1983. Edited by H. Araki, C Moore, Ş. Strătilă and C. Voiculescu. VI, 594 pages. 1985.

Vol. 1133: K.C. Kiwiel, Methods of Descent for Nondifferentiable O mization. VI, 362 pages. 1985.

Vol. 1134: G.P. Galdi, S. Rionero, Weighted Energy Methods in F Dynamics and Elasticity. VII, 126 pages. 1985.

Vol. 1135: Number Theory, New York 1983–84. Seminar. Editec D.V. Chudnovsky, G.V. Chudnovsky, H. Cohn and M.B. Nathans V, 283 pages. 1985.

Vol. 1136: Quantum Probability and Applications II. Proceedir 1984. Edited by L. Accardi and W. von Waldenfels. VI, 534 pag 1985.

Vol. 1137: Xiao G., Surfaces fibrées en courbes de genre deux. IX, pages. 1985.

Vol. 1138: A. Ocneanu, Actions of Discrete Amenable Groups on Neumann Algebras. V, 115 pages. 1985.

Vol. 1139: Differential Geometric Methods in Mathematical Phys Proceedings, 1983. Edited by H. D. Doebner and J. D. Hennig. VI, pages. 1985.

Vol. 1140: S. Donkin, Rational Representations of Algebraic Grou VII, 254 pages. 1985.

Vol. 1141: Recursion Theory Week. Proceedings, 1984. Edited H.-D. Ebbinghaus, G.H. Müller and G.E. Sacks. IX, 418 pages. 19

Vol. 1142: Orders and their Applications. Proceedings, 1984. Ec by I. Reiner and K. W. Roggenkamp. X, 306 pages. 1985.

Vol. 1143: A. Krieg, Modular Forms on Half-Spaces of Quaternic XIII, 203 pages. 1985.

Vol. 1144: Knot Theory and Manifolds. Proceedings, 1983. Editec D. Rolfsen. V, 163 pages. 1985.